Australia's Dangerous Creatures

FOR

DUMMIES®

by Graeme Lofts and Peg Gill

WILEY

Wiley Publishing Australia Pty Ltd

Australia's Dangerous Creatures For Dummies®

published by
Wiley Publishing Australia Pty Ltd
42 McDougall Street
Milton, Qld 4064
www.dummies.com

Copyright © 2008 by Wiley Publishing Australia Pty Ltd

The moral rights of the authors have been asserted.

National Library of Australia
Cataloguing-in-Publication data

Lofts, Graeme
Gill, Peg

    Australia's dangerous creatures for dummies.

    Includes index.
    ISBN 978 0 731 40722 4 (pbk).

    Dangerous animals — Australia.
    Poisonous animals — Australia.

591.650994

Cover image: Getty Images/National Geographic/Jason Edwards

Cartography by MAPgraphics Pty Ltd, Brisbane, and the Wiley Art Studio

10 9 8 7 6 5 4 3 2 1

About the Authors

Graeme Lofts is the author of more than 25 science and physics textbooks for Australian secondary schools, including the award-winning Science Quest series. He has also written and produced videos for schools on topics ranging from the physics of sound and light, to the preservation of Australian animal habitats.

After graduating from the University of Melbourne with a major in Physics, he took up teaching Physics, but soon found himself increasingly drawn to other areas of science, including environmental science and geology. After teaching and lecturing in Australian schools and universities for 22 years, he decided to concentrate on writing.

Graeme has travelled extensively throughout Australia and enjoys bushwalking and camping. His favourite locations are the Otway Range of Victoria and the tropical rainforests of northern Queensland. He lives with his wife, Dianne, in Sunbury, on the outskirts of Melbourne.

Peg Gill is a graduate of the University of Melbourne, where she majored in Genetics, Zoology and Botany. After 30 years of teaching and lecturing in Biology, she has retired to pursue a career in writing. Her love of nature and enthusiasm for learning about plants and animals continues to infuse her daily living.

Peg, along with her husband, Ian, and three children, lived for many years in the small country town of Riddells Creek, Victoria, in the foothills of the Macedon Ranges. Here, bushwalking and camping could be easily enjoyed and animals from the bush often visited.

In 2007 she swapped the bush for the beach and now lives in the seaside suburb of Black Rock, Victoria, where she loves to explore the beaches and rock pools of Port Phillip Bay.

Peg and Graeme have been friends since studying at university together, more than 35 years ago.

Dedication

We dedicate this book to those who work with a passion for the conservation of all creatures great and small, and those with the adventurous spirit that drives them to explore the diversity of Australia and its wildlife.

Authors' Acknowledgements

This book would not have been possible without the helpful support of our spouses: Dianne Lofts, who found numerous informative Web sites that we could not have found by ourselves; and Dr Ian Gill, who provided critical and constructive advice on first aid.

We would also like to acknowledge the professional and friendly support of Charlotte Duff at Wiley, who offered much helpful advice, and our editor Maryanne Phillips and US review editor, Jennifer Bingham. Thanks also to the artists who created the images that we requested and worked creatively with our suggestions for cartoons.

Thankyou to those who provided expert advice and statistics along the way: Dr Ken Winkel (Australian Venom Research Unit, www.avru.org), Dr Karen Ashby (Victorian Injury Surveillance Unit), Russell Hore (Quicksilver Connections, Port Douglas), Ray Lord (Royal Society for the Prevention of Cruelty to Animals Victoria), Charlie Manolis (Wildlife Management International), Katie Maynes (Australia Zoo), Michael McFadyen (www.michaelmcfadyenscuba.info), Robert van der Hoek (Australian Institute of Health and Welfare), John West (Curator, Australian Shark Attack File, Taronga Zoo, Sydney), the Aquarium of Western Australia (AQWA), the Australian Museum and the National Coroner's Information System. The CSIRO Division of Entomology (www.csiro.au/org/Entomology) inadvertently provided support through its magnificent Web site.

Publisher's Acknowledgements

We're proud of this book; please send us your comments through our Dummies online registration form located at www.dummies.com/register/.

Some of the people who helped bring this book to market include the following:

Acquisitions, Editorial and Media Development

Project Editor: Maryanne Phillips

Acquisitions Editor: Charlotte Duff

Technical Reviewer: Dr Ken Winkel

Editorial Manager: Gabrielle Brady

Production

Layout and Graphics: Wiley Composition Services and the Wiley Art Studio

Illustrations: Graeme Tavendale

Cartoons: Glenn Lumsden

Proofreader: Amanda Morgan

Indexer: Don Jordan

The authors and publisher would like to thank the following copyright holders, organisations and individuals for their permission to reproduce copyright material in this book.

Black & white images: MAPgraphics Pty Ltd, Brisbane: page 12 • Australian Venom Research Unit: pages 64, 66, 69, 70 and 146/distribution maps sourced from the Australian Venom Research Unit, Department of Pharmacology, University of Melbourne, Victoria, Australia (www.avru.org) • © Viewfinder Australia Photo Library: pages 92, 107 (fruit bat), 134, and 168 (plover) • ANTPhoto.com.au: page 98 /Otto Rogge; page 107 (insectivorous bat) /Karin Cianelli; page 121 /Martin Willis; page 163 /Ron & Valerie Taylor; page 231 /Ian McCann; page 240 /Cyril Webster • © EyeWire Images: page 102 /cat) • © Photodisc: pages 102 (dog) and page 168 (silvergull) • © Image Disk Photography: page 110 (2 images) and page 168 (pelican) • Corbis Australia: page 152 /Gary Bell/zefa; page 158 /Stephen Frink; 168 (tern) /Pam Gardner/Frank Lane Picture Agency • © CSIRO Entomology: pages 154 and 239 • © Purestock Superstock: page 157 • Photolibrary.com: page 178 (Black Sea Urchin) /Science Photo Library/Alexis Rosenfeld; page 202 /Dave Fleethman; page 233 /Peter Arnold Images Inc./Ed Reschke • Corbis Royalty Free: page 178 (Crown of Thorns Starfish) © Corbis Digital Stock • © Digital Vision: pages 197 and 208 /Stephen Frink; page 244 • © Digital Stock/Corbis Corporation: page 203 • Newspix: page 209 /Bob Fenney; page 259 • © Brand X Pictures: page 228 • Tasmanian Museum and Art Gallery: page 261

Colour images: Corbis Australia: page 1 (Saltwater Crocodile) /David A. Northcott; page 5 (octopus) /Gary Bell/zefa; page 6 (stonefish) /Jeffrey L. Rotman; page 8 (wasp) /Joe McDonald • ANTPhoto.com.au: page 1 (brown snake) and page 2 (tiger snake) /Rob Valentic; page 2 (taipan) and page 3 (tick) /Michael Cermak; page 3 (box jellyfish) /Kelvin Aitken; page 5 (cone shell) /Ron & Valerie Taylor; page 8 (bull ant) /Cyril Webster • Newspix: pages 4 (Irukandji) and 6 (Red-back Spider); page 7 (funnel-web spider) /Sam Ruttyn • Corbis Royalty Free: page 4 (shark) /© Marty Snyderman/Corbis Digital Stock • © Brand X Pictures: page 7 (bee)

Text: Taronga Zoo: page 156 (table 10-1) /Australian Shark Attack File, Taronga Zoo, Mosman, Sydney

Every effort has been made to trace the ownership of copyright material. Information that will enable the publisher to rectify any error or omission in subsequent editions will be welcome. In such cases, please contact the Permissions Section of John Wiley & Sons Australia, Ltd who will arrange for the payment of the usual fee.

Contents at a Glance

Table of Contents

Introduction

. . . .

*A*ustralia has more than a fair share of dangerous animals. Some of them, like crocodiles and sharks, are downright deadly. Many others are capable of causing serious injuries or illness.

Still, every animal on this planet, no matter how dangerous, is a 'creature of beauty' and has an important role to play in keeping a balance in nature. In Australia, you're never far from dangerous creatures, whether you're at home or enjoying the great outdoors. We hope that you can appreciate them without putting yourself at risk.

About This Book

In this book, we aim to provide you with enough knowledge to help you recognise some of the world's most dangerous animals but still avoid unpleasant encounters with them. We also suggest what to do if you're attacked.

The variety of creatures that live in Australia is matched by an amazing diversity of scenery and habitats. In *Australia's Dangerous Creatures For Dummies* we take you on a tour overland to show you creatures that live in tropical rainforests, woodlands, swamps, open plains, deserts and snow-covered mountains. We also tour the coast, to uncover the creatures that live in and around rocky shores, sandy beaches, coral reefs and the open sea. You even discover the suburban dwellers — animals living in or visiting suburban parks, gardens and even bedrooms in homes.

Conventions Used in This Book

In this book, we try to avoid using medical jargon when describing the effects of attacks by dangerous creatures. However, where medical terms provide the most concise way of describing injuries or symptoms, they are used. In these instances we explain the terms.

We also use the common names of animals rather than scientific names. The names of individual species of animals begin with upper case (for example, Great White Shark and Sydney Funnel-web Spider). Upper case is not used in the names that describe groups of animals (for example, the death adder snake — yes, more than one species of this snake exists).

What You're Not to Read

You don't need to read this book from cover to cover — just read what interests you or the sections on the destinations you're likely to visit.

Sidebars are shaded boxes scattered throughout the book. Feel free to skip them. The sidebars contain information that's not essential to understanding topics, but go into more depth or add a bit of interest.

You can also skip the paragraphs beside the Technical Stuff icon. And if you've lived in Australia for any length of time, you probably won't need to read the paragraphs beside the Australiana icon. (For information about all the icons we use, flip over to 'Icons Used in This Book' later in this Introduction.)

Foolish Assumptions

In writing this book we have assumed that you want to know more than you already do about Australia's dangerous creatures for one or more of the following reasons:

- ✔ You just love learning about creatures
- ✔ You have an interest in Australia

✔ You intend travelling in Australia

✔ You enjoy bushwalking or swimming

✔ You would like to remain safe and healthy

We've also assumed that:

✔ You have an average general knowledge of animals, but little or no scientific knowledge of their physiology or classification.

✔ You have little or no knowledge of medicine or first aid.

How This Book Is Organised

This book is divided into six parts. Here's a run-down on what you can find in each of them.

Part 1: Getting Acquainted with Australia's Creatures

This part tells you why Australia's dangerous animals are hazardous to humans and where you're likely to encounter them. It explains how to avoid dangerous encounters when you're on the move — whether on foot or on wheels — and when you're camping. Also, this part explains how, with planning and preparation, you can reduce the chance of encounters with dangerous creatures and what to stock in a first aid kit in case of an emergency.

Part II: Exploring the Great Southern Land

The variety of dangerous creatures on Australian soil is amazing. This part describes the creatures that you could encounter inland, explains how you can avoid being harmed by them and how to apply first aid in the event of a nasty encounter. You find out about creatures as large as crocodiles, as unique as the Platypus or the Tasmanian Devil, and as tiny as mosquitoes, ticks and many more.

Part III: Bays and Beaches

This part takes you to the coast, where you could meet deadly sharks, lethal jellyfish, spiky stinging fish and other dangerous, but beautiful creatures at sandy beaches, in rock pools, around coral reefs or further out to sea. You find out how to avoid these marine creatures and how to treat stings or bites — in some cases while waiting for an ambulance.

Part IV: Urban Living

You don't have to travel far to find dangerous, or even deadly, creatures. Some of them are quite at home in your own backyard, in urban parks and gardens, and even indoors. This part explains the hazards of spiders, bees, wasps and other tiny creatures that can make you very ill — or even kill you. You also find out about some slightly larger creatures — like rats, mice and even birds — that can do you harm.

Part V: The Part of Tens

In this part, you can find our list of the ten most deadly creatures found in Australia (such lists are, of course, a matter of opinion). We also make ten suggestions about how to stay safe from dangerous creatures and list ten places where you can safely see Australia's dangerous creatures 'up close and personal'.

Part VI: The Appendixes

Although first aid procedures following animal attacks are described throughout this book, we want you to be able to locate first aid information as quickly as possible in the event of an emergency. We include two appendixes on first aid — the first describing emergency techniques and the second listing remedies by creatures' names — so that you can find the information you need in one place, rather than having to flick through the whole book or use the index.

Icons Used in This Book

You may notice icons, or neat little pictures, in the margins throughout this book. These icons do more than just break up the white space — they tell you something about the particular paragraph where they appear.

This icon highlights useful hints that can keep you safe from danger.

The Remember icon indicates important points that we think you need to commit to memory for later access.

This icon alerts you to something that you should *not* do if you want to avoid nasty (or deadly) encounters.

It's hard to resist getting into some of the science of dangerous creatures — it's so fascinating. The Technical Stuff icon indicates that the information is a little scientific — and can be skipped unless you're really interested.

This icon briefly describes first aid procedures in the event of a bite, sting or other injury sustained in an encounter with a dangerous creature.

Just in case you're not familiar with some of the local terminology, this icon highlights terms that are uniquely Australian.

Where to Go from Here

Chapter 1 is a great place to start, especially if you don't live in Australia — you can discover what makes creatures dangerous. Chapters 2 and 3 provide useful information if you want to explore Australia, and be safe while doing so.

From there you can pretty much go anywhere you want. Like any tour, you can choose which places you'd like to visit and in which order. If you're travelling inland, Part II is the place to go. If you're heading to the beach or offshore, turn to Part III. Or if you're content to stay at home — or just visit the cities, suburbs and towns — Part IV reveals the many creatures to be wary of.

Whatever route you take, this book provides plenty of information about dangerous creatures, how to avoid them and some basic information about first aid in the event of an encounter.

If it's just first aid information that you need, skip to the appendixes.

Part I
Getting Acquainted with Australia's Creatures

Glenn Lumsden

'You're meant to take the wrapper off them first, Bert.'

In this part . . .

We'd like to introduce you to the Australian countryside and its coastal waters so that you can enjoy the trip of a lifetime or simply a great day's outing.

In this part, we provide you with some advice on how to plan your journey, how to avoid encounters with Australia's dangerous creatures, and what to take with you so that you come home safely.

Chapter 1

Revealing the Dangers of Australian Wildlife

*A*ustralia is home to some of the most dangerous animals in the world. Some species, including the ten most venomous snakes on earth, are not found in any other part of the world (well, except in zoos). And Australia has a huge variety of other dangerous creatures — from giant sharks and crocodiles that can literally tear you apart, to venomous spiders, cute looking Koalas with very sharp claws, to tiny but nasty ticks and fleas that can spread deadly diseases.

So, whether you're an Aussie at home or on holidays, or a visitor from another country, you need to know how to avoid accidental encounters with dangerous animals. You also need to know what to do if you do come face to face with one and how to respond if you're bitten, stung or otherwise injured. In this chapter, we lay some ground rules.

The Nature of the Danger

Dangerous doesn't necessarily mean deadly. We have taken the view that dangerous means capable of causing injury or illness either directly (for example, due to venom or blood loss) or indirectly (as a result of infection of a wound, allergic reaction, transferring a disease or perhaps causing a car accident).

Where the danger lies

Animals dangerous to humans usually have one of these four characteristics:

- ✓ **Venom that can be injected into the bloodstream.** *Venom* is a substance that contains one or more toxins that can cause tissue damage, such as blood clotting, cell death, muscle damage and nerve damage. Venom is injected into the victim with fangs, spines, other pointy structures or stinging cells. Some venoms are deadly — capable of killing a human within minutes. Others cause little more than minor swelling and irritation. Fortunately, *antivenoms*, which neutralise the effects of most venoms, are now available; fatalities caused by venomous creatures are less likely than they were 50 years ago.

- ✓ **Sharp teeth and powerful jaws that can rip into human flesh, causing serious injuries.** Crocodiles use their jaws and teeth to crush their prey, whereas other creatures, such as sharks, use their teeth to tear flesh apart. Most meat-eating mammals, even small ones like rats and mice, have teeth that can cause injuries to humans.

- ✓ **Claws that cause cuts and scratches that can later become infected.** Crabs, Emus and Koalas are just some of the animals with dangerous claws. The giant Wedge-tail Eagle has claws that are sharp enough to penetrate the vital organs of its normal prey — and could potentially do the same to a small toddler. Even centipedes inject venom through their claws.

- ✓ **Venomous secretions that can harm the skin or the eyes of victims.** Cane Toads spray venom from glands around their neck. Millipedes release a toxic substance from their bodies that stains and burns.

Other, less common characteristics that make creatures dangerous to humans are:

- ✔ **Sheer size.** Imagine the effects of being knocked over or trampled by a 1,200-kilogram (2,640-pound) water buffalo or a rampaging feral horse, donkey or pig.
- ✔ **Poisonous flesh.** Although the pufferfish isn't venomous, its flesh contains a deadly toxin. If this fish isn't prepared and cooked correctly, eating it can cause headaches, vomiting, internal bleeding, paralysis and even death. Some turtles and crabs also have poisonous flesh.

Allergies to Venom

Unfortunately, many people suffer allergic reactions to some venoms, particularly bee, wasp and ant venom. The reactions might involve a severe rash, swelling, stomach cramps, diarrhoea and vomiting. The most severe allergic reaction is known as anaphylaxis, which can be fatal. *Anaphylaxis* (or anaphylactic shock) is a sudden, severe allergic reaction to a substance. It requires immediate medical attention (as we note in Appendix A). Symptoms include breathing difficulties, swelling (particularly of the face, throat, lips and tongue in cases of food allergies), rapid drop in blood pressure, dizziness, unconsciousness, hives, tightness of the throat and a hoarse voice.

Anaphylaxis can happen in just seconds after being exposed to an allergic substance or can be delayed for up to two hours if the reaction is from a food.

Getting Your Bearings

When you're travelling, you probably want to know where you're likely to encounter dangerous creatures. That means you need information about their natural habitats, and which regions you might find them in. This is easier if you get your bearings first. The map shown in Figure 1-1 can help.

The term *Top End* is used to describe the far north of Western Australia, the Northern Territory and Queensland. The *Outback* is the arid central part of Australia. Most of the Australian mainland could be described as outback.

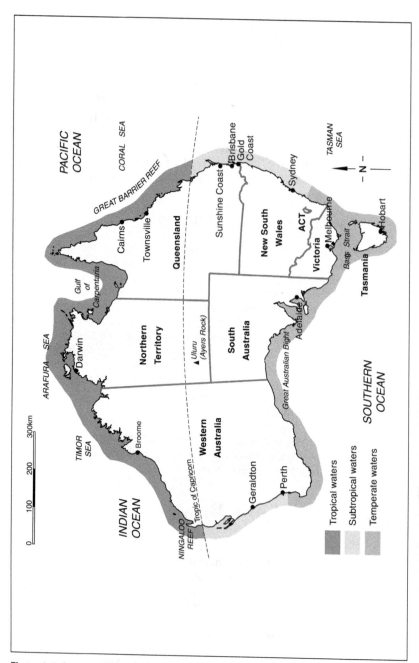

Figure 1-1: Australia: The great southern land.

Danger Zones

To give you a general idea of where you might find Australia's dangerous creatures, here's a brief summary outlining where they live.

On the ground

Almost everywhere outdoors in Australia is inhabited by creatures. You could encounter snakes, Emus, echidnas, kangaroos and feral animals. Although some of the snakes are deadly, the venom of most will just make you quite ill. Emus, echidnas, kangaroos and feral animals are unlikely to kill you, but they can cause nasty wounds. And any mammal could be carrying ticks, which are dangerous in their own right.

When you're out walking, keep your hands and face away from the hollows of fallen trees — they're great places for all sorts of creatures to shelter, including snakes.

Hiding under a rock

You shouldn't need to move rocks when you're out in the bush or at the beach. Apart from the small but dangerous creatures you might disturb, you're interfering with the habitat of numerous other small creatures. But if you must move a rock, wear a sturdy pair of gloves and watch out for creatures like spiders, scorpions and centipedes. Along Australia's rocky shores you need to watch out for creatures like crabs, cone shells and deadly blue-ringed octopuses if you disturb rocks.

Australians call everywhere outside capital cities and large regional cities and towns the *bush* — with the possible exception of the sandy deserts.

Among the trees

When you're in woodland or a forest you need to keep your eyes open for snakes and any small creatures that might shelter in the hollows of tree trunks. Many creatures build their nests in trees, including bees, wasps, eagles and the

Magpie. They defend their nests and young vigorously, and serious injuries can be inflicted on anyone who dares to get too close. And if you throw an object at a nest, you're asking for trouble.

Koalas are active at night and might be moving from one tree to another — best to keep right out of their way to avoid some nasty scratches. Watch out for leeches in damp forests. If you're lucky enough to spot a Southern Cassowary in the tropical rainforests of northern Queensland, observe quietly from a distance to avoid a powerful kick.

Watch the water

Swimming, snorkelling, diving and even wading in Australian waters, whether at the beach, in and around coral reefs or in rivers and creeks, can be hazardous. In tropical waters (beaches and rivers), you have to be aware of the possible presence of crocodiles and deadly jellyfish. Everywhere else you could come across sharks, stingrays, sea snakes and spiky fish capable of killing a human with their venom. In creeks and rivers, even the harmless-looking Platypus is dangerous.

Around the home

Most of the creatures mentioned in this chapter can be found in and around Australia's homes and gardens. Which ones, of course, depend on the location. You can encounter snakes (including the deadly ones), spiders (like the deadly Red-back and Sydney Funnel-web), bees, wasps and ants — especially during the warmer months. Even in the apparent safety of your own bedroom lurk smaller creatures, such as bedbugs and fleas, which can spread nasty diseases.

A Case of Mutual Respect

Almost all of Australia's dangerous creatures are unlikely to attack humans unless they're provoked. The provocation can be deliberate (like disturbing a bees' nest or chasing a snake) or accidental — like sitting on a Red-back Spider that you haven't seen or stepping on a stingray half buried in a sandy beach floor. Almost all attacks can be prevented by behaving

Humans: The most dangerous creatures of all

The most dangerous creatures in Australia, and throughout the rest of the world, are humans. Populations of crocodiles, sharks, lions and tigers, to name just a few, have dramatically decreased over the past 200 years as a result of human activity. The removal of natural habitats for urbanisation and hunting are the biggest threats to creatures. Some species, such as the Tasmanian Tiger, have become extinct since European settlement, and others, such as the Grey Nurse Shark and the Southern Cassowary, are under serious threat of extinction. Saltwater Crocodiles were hunted almost to extinction until they became protected by law more than 30 years ago.

Humans have introduced species from other parts of the world that end up competing with or preying on Australia's native creatures, thereby threatening their survival. For example, introduced rabbits, sheep and cattle compete for food with kangaroos and wallabies. Domestic pets, like cats and dogs, can become dangerous feral pests, hunting and competing with native creatures. And man's misuse of waterways has grossly affected water quality, flow rates, sediment and temperature, endangering the survival of native fish and other aquatic creatures.

sensibly and being alert when you're engaging in outdoor (and some indoor) activities. Co-existing with other creatures is really a case of mutual respect.

Keeping Your Safety in Mind

You don't have to go on a trip to be in danger from an animal bite or sting. Even at home you need to be prepared for visits from dangerous creatures.

Throughout this book, we discuss many fascinating Australian creatures, but we don't intend to frighten you. We want you to observe and delight in Australia's creatures — dangerous or otherwise. Animals aren't dangerous if you know how to act appropriately and you treat them with respect. This book tries to help you with these encounters.

Housekeeping

To ensure that your home is not a haven for dangerous creatures, place rubbish and food scraps outside the house in secure garbage bins so that you don't attract wasps, ants and flies, or even larger scavengers like feral cats, dogs and pigs.

Water bowls and stagnant pools need to be emptied and cleaned to prevent mosquitoes (and Cane Toads in some parts of Australia) from breeding.

Wood-piles, old bricks and rubble provide ideal homes for snakes, lizards and spiders. These need to be tidied and stacked away from the house and away from where children play. Keep outside sheds clean and tidy to discourage spiders.

Gardens need to be turned over to disrupt bull ant and termite nests. (Make sure you've got sturdy boots on while you do this.) And cut long grass, so that it can't provide shelter for snakes.

Inside the house, regularly clean under beds and other furniture, wash blankets and doonas, and steam clean the carpets and rugs to get rid of mites and lice. Go through the cupboards and throw out stale and unwanted packets of food to deter mice and rats. Apart from the fact that mice and rats can spread disease or bite you, they also attract snakes.

First aid

Everyone needs a first aid kit at home or when they travel. We suggest a list of items to include in your kit in Chapter 3. The contents of a kit need to be checked regularly for three reasons:

- To ensure that the items are clean, the packaging is still sealed and sterile, and that expiry dates are still valid
- To check that items used are replaced
- So that you remain familiar with what's in your kit, where each item is and how to use them

We include first aid instructions throughout this book — after descriptions of dangerous creatures and in the appendixes.

Chapter 2

Venturing Out

*C*hoices, choices, choices. You have so many different ways to explore this country — Australia is such a diverse land. You can drive a car, ride a bike or trek on foot through tropical or temperate rainforests, barren plains and deserts, snow-covered alpine slopes or coastal parks overlooking sandy bays or rocky reefs. You can explore for a single day or absorb and enjoy the surroundings over an extended period of time. You can camp close to nature, or if you prefer more luxurious surroundings, choose from a range of more comfortable accommodation wherever you go.

No matter where you go, though, you should be prepared for encounters with wildlife. And each method of getting around presents its own hazards. In this chapter, we explore ways to safely step out and experience the great outdoors.

Bushwalking Paradise

Australia is a bushwalker's paradise. Most bushwalking trails begin only a few hours' drive from our cities. A walk may be a pleasant half-hour stroll in the bush, or a trek overland taking several weeks. You can experience mountains, rainforest, deserts and even Antarctic conditions. The terrain and vegetation in Australia's national parks, state forests and reserves vary greatly, providing something for everyone.

Bushwalking can be attempted all year round, with two main exceptions:

- ✔ Parts of the Top End (northern Queensland, Western Australia and the Northern Territory) can be uncomfortably hot and wet in summer.
- ✔ Much of the Australian Alps are snowbound during winter.

Everywhere else is open for trekking, provided you're well equipped for the conditions. Some well-known trails need to be booked early, though, because accommodation and other facilities can be limited. (Later in this chapter, we provide resources so that you can make bookings.)

Australia is relatively safe for backpackers and hikers. Cheap, clean accommodation is available in all large cities, and good camping sites and sometimes huts or cabins are available on the trails. The native animals may not be all that friendly, but most are timid and will leave you alone, if you leave them alone. And none is likely to attack you at night in your tent!

Dealing with wild encounters

Bushwalking allows you to take in Australia's wonderful scenery, but may also bring you 'up close and personal' with some of Australia's wild creatures — sometimes a little closer than you would like.

You're safer letting native animals know you're approaching — inadvertently sneaking up on them may frighten them and cause them to react aggressively. When you're walking make *some* noise; a steady tread and a little conversation warns animals that you're around. Most snakes and many other potentially dangerous creatures will retreat if they hear you coming. To observe animals, keep your distance. Look around for a nice safe spot to rest quietly, and the wildlife will soon reappear.

Never put your hands in leaf litter, up a hollow log or in a hole — you might get bitten and you may not know what bit you, making first aid difficult. Also, check your clothing before starting off again; you can easily pick up a spider, tick, leech or small reptile while you rest.

Organising a bushwalk

Planning ahead is essential when you're bushwalking. Self-sufficiency on a bush walk is vital. No one plans on getting lost or wishes to get locked in by weather conditions, but the possibility is always there. So, like every Scout or Girl Guide, 'be prepared'.

By contacting the national park or state forest that oversees the walk you're interested in, you can obtain details about accommodation and camping facilities, the walk's degree of difficulty, alternative bush trails, best times to visit, dangerous creatures you could encounter, emergency contact phone numbers and other relevant information. The resources listed in Table 2-1 make great starting points, and you can use them to find out about particular bush walks.

A little respect goes a long way

Native bushland, forests and beaches are places to enjoy, discover and experience what nature has to offer. So please respect and don't tamper with animals, vegetation, Aboriginal sites and rock formations in any way.

When bushwalking, carry your litter out with you.

If no toilets are provided, bury all waste at least 100 metres (110 yards) from water courses to avoid contaminating the environment.

Table 2-1	Planning a Bushwalk	
State or Territory	*Name of Resource*	*Web Address*
Australian Capital Territory	ACT Department of Territory and Municipal Services (Environment)	www.tams.act.gov.au/ live/environment
New South Wales	NSW National Parks and Wildlife Service	www.nationalparks. nsw.gov.au/npws.nsf/ Content/Home
	NSW Department of Primary Industries	www.dpi.nsw.gov.au/ forests
Northern Territory	Natural Resources, Environment and the Arts (Parks)	www.nt.gov.au/nreta/ parks
Queensland	Queensland National Parks and Wildlife Service	www.epa.qld.gov.au
South Australia	Department for Environment and Heritage	www.dehaa.sa.gov.au
Tasmania	Parks and Wildlife Service	www.parks.tas.gov.au/ tpws.html
Victoria	Parks Victoria	www.parkweb.vic.gov.au
Western Australia	Department of Environment and Conservation NatureBase	www.calm.wa.gov.au

The clothing you need to carry depends on the time of year, the length of the trail and the area that you're hiking in. Be it a short walk or a several-day trek, be prepared for all weather conditions.

A checklist of clothing and equipment to consider taking with you is covered in Chapter 3, but when bushwalking always carry:

- ✔ Plenty of water, especially in the warmer months
- ✔ Some food (snacks, fruit or meals)
- ✔ A hat (to keep the sun off your face in summer or to keep your head warm in winter)
- ✔ Sturdy walking shoes or boots
- ✔ Sunscreen
- ✔ A good map of the area
- ✔ A compass — a good addition to any map, but make sure you know how to use it
- ✔ A mobile (cell) phone if you have one, to call for help if the need arises
- ✔ A mirror (in case the phone doesn't work — with a mirror you can 'signal' your presence to a search party)
- ✔ A first aid kit

Check the weather report before leaving and take a suitable jacket, just in case rain or cold sets in. If your walk involves overnight stays, ensure you have enough warm clothing and blankets or a sleeping bag to suit the conditions. Even if the weather's hot, wear long trousers and a long-sleeved shirt to protect yourself from bites, particularly from snakes.

Always inform someone of your plans so that they can raise the alarm if you don't return at the expected time. Tell someone at the place you're staying, a friend or family member, or at the office of the park from which you're leaving. Many parks have a trekker's 'sign in and out' book.

Be aware of bushfire risk, especially during the bushfire season — September to April. Only light fires in fireplaces provided and observe the total fire ban days.

Join the (bushwalking) club

We recommend you never walk alone. So, if you're travelling by yourself, join an organised hike.

Bushwalking or hiking clubs are located all over Australia. Many provide social evenings and the opportunity to become involved in a local project, as well as join organised walks. These clubs are also a great way to find like-minded people and provide a safe way to bushwalk in unknown terrains. Walks are scaled to various degrees of difficulty, so something exists for everyone.

The easiest way to track down a club in the region you're visiting is to jump on the Internet or ask a local Tourist Information Centre. From there, you can attend a meeting or make a phone call and decide whether you want to join the club in that area. Club fees vary, according to the activities they undertake, but usually cost around $20 to $40 per year.

National parks may be closed to visitors at times of high fire danger. Warnings about fire restrictions are on radio and television news reports. You can also find out about fire warnings in all states and territories by checking the Bureau of Meteorology Web site (www.bom.gov.au).

 Bushfire is an Australian term for forest fires or wildfires. Throughout Australia, Fire Control Authorities declare *total fire ban* days when the weather and dry vegetation create a high risk of bushfire. These bans may be statewide or confined to particular regions of a state. On total fire ban days, no fire may be lit in the open.

On the Road

Unless you're prepared to take a plane, bus or train, the best way to get from place to place in Australia is to drive. Good all-weather bitumen roads are found all around Australia, and well-maintained highways and motorways link major cities. You can get to most places in Australia in a typical suburban car, but you need a four-wheel-drive for non-bitumen roads that are off the beaten track.

Encountering wildlife on the road

Don't believe for one minute that wildlife can't endanger you when you're inside a motor vehicle. Road fatalities caused by motorists colliding with animals or trying to avoid animals crossing the road are quite common.

Accidents involving motor vehicles and animals are most common at dawn, dusk and night-time, when many of Australia's native creatures are at their most active. This problem isn't confined to the outback or rural areas. Kangaroos, for example, present a danger on the outskirts of most Australian cities and towns.

Avoiding collisions

Road signs indicate stretches of road where animals are known to cross, so take heed of these warnings, and slow down between dawn and dusk. When travelling at the speed limit, avoid swerving to miss an animal; you're more likely to have a serious accident with a tree. If you see an animal on or approaching the road, sound your horn in a series of short blasts — this usually frightens the animal away. (For tips on avoiding collisions when riding on two wheels, see the section 'Enjoying the ride without hitting the wildlife' later in this chapter.)

Driving Australia

If you hold a driver's licence issued outside Australia, you need to be aware of the road rules. In particular, you need to know that in Australia:

✔ Vehicles are driven on the left side of the road.

✔ Seat belts are compulsory, whether you're a driver or a passenger.

✔ It's illegal to talk on a mobile (cell) phone while driving.

✔ Speed limits are strictly enforced.

✔ Strict drink-driving laws apply. 'P' plate drivers can't consume alcohol prior to driving (a zero blood alcohol reading), and experienced drivers can't have a blood alcohol reading above .05.

To reduce your impact on wildlife, consider these driving tips:

✓ Avoid driving at dusk and dawn in areas populated by wildlife. This is when animals come out to feed and are more likely to be near or crossing roads.

✓ Be alert, whether you're the driver or a passenger. Road signs often indicate where animals are likely to be found. Use these signs as guides to be on the lookout for animals on and at the sides of roads.

✓ Drive cautiously through these areas. Reduce your speed and use your high-beam lights where possible to enhance your vision.

✓ Notify authorities if you see an injured animal or are involved in injuring an animal. Local police stations or a motoring organisation (see the related sidebar) can help.

✓ Move dead animals well away from the side of the road, if doing so is safe. If the animal is a marsupial (such as a kangaroo, wallaby or Koala) check the pouch for any surviving young. Don't try to remove the survivor, though; notify local authorities.

Avoiding unwanted hitchhikers

When you stop to take a break from driving or to have a closer look at the sights, make sure that you don't accidentally pick up a wildlife hitchhiker. Insects, spiders and other small creatures can easily get caught up in a jacket or rug that you've put on the ground when you've stopped for a walk, rest or picnic. Give your gear a good shake before you put it back in the car. Even snakes can get caught up in rugs or find their way into picnic baskets.

Crossing rivers and creeks safely

Crossing a river or creek with no bridge when driving in remote areas presents its own special challenges. First you need to determine if your vehicle can cross. If you're not in crocodile country, get out of the car and try to wade across (with care — we don't want you swept way!). Use a stick in front of you to determine the depth and speed of the water. But don't wade in bare feet — wear a pair of sturdy boots. Apart from the fact that you might stand on something sharp, or that you could accidentally stand on some helpless creature, venomous spiky fish live in some rivers and creeks. Only cross if you're certain it's safe.

Another way to determine the crossing's depth is to watch other motorists — if they're around. Before crossing, observe someone else attempting the river or creek and be guided by how deep they find the water to be.

Never allow the water level to reach the top of your wheels because the vehicle may be washed away.

Motoring organisations

Each Australian state or territory has its own motoring organisation, and all of them belong to the Australian Automobile Association (AAA). They can provide a wealth of information about regions in their state and have reciprocal arrangements to aid inter-state travellers. Some provide free maps, guides and accommodation information, and if your car breaks down or is involved in an accident they can send someone to your aid. They can also tell you what to do for an injured animal and where the closest help or shelter is.

Roadside assistance can be contacted by calling 13 11 11 in any state.

For touring and travel information before you leave, or when on holiday, contact the motoring organisation in the state or territory:

Queensland: RACQ
300 St Pauls Terrace, Fortitude Valley
Tel (07) 3361 2444
Web site: www.racq.com.au

New South Wales: NRMA
74–76 King Street, Sydney
Tel 13 21 32
Web site: www.nrma.com.au

Victoria: RACV
360 Bourke Street, Melbourne
Tel 13 19 55
Web site: www.racv.com.au

Tasmania: RACT
Corner Murray & Patrick Streets, Hobart
Tel 13 11 11
Web site: www.ract.com.au

South Australia: RAA
41 Hindmarsh Square, Adelaide
Tel (08) 8208 4600
Web site: www.raa.com.au

Western Australia: RAC of WA
228 Adelaide Terrace, Perth
Tel (08) 9421 4444
Web site: www.racwa.com.au

Northern Territory: AANT
81 Smith Street, Darwin
Tel (08) 8981 3837
Web site: www.aant.com.au

On Your Bike

You can get to some great places that cars can't access by pedalling a bike or riding a motorbike. And the feeling of the wind on your face and the scent of the eucalyptus in the air means you're just that little bit closer to nature than in a car. In Australia, though, you have to wear a bike helmet and obey the same road rules as motorists.

Push bikes

Cycling is popular in Australia. Lots of bike paths and trails are designed especially for riders, although you need to be aware that many paths are shared with pedestrians and in-line skaters. Tourist information centres carry information about bike routes and rentals. Many organisations also run bike tours. In addition, you can often ride from one town to another on quiet back roads devoid of traffic. You can also take your bike on most trains and ferries (but not at peak hours: Usually 8.00–10.00 am and 5.00–7.00 pm) to get it from one major destination to the next. Some rail and ferry lines may charge extra to carry your bike.

Tours on wheels

Joining an organised tour that offers expert support, specialised knowledge of an area, accommodation and meals and the opportunity to see places others never get to, can be a wonderful experience. Travelling this way can also be safer. Bicycle and motorbike tours are based in all major cities and head to every type of destination — rainforests, mountains, deserts, beaches, and even Uluru.

Motorbikes

Australia is a fantastic place to ride a motorbike. Some of the best motorcycle routes in the world include Cape York in Queensland; the Blue Mountains and the Hunter Valley in New South Wales; the Great Ocean Road along the Victorian coastline; the Adelaide Hills in South Australia; the Kimberley Ranges in Western Australia; and the northwest of Tasmania.

Enjoying the ride without hitting the wildlife

Whether you use pedal-power or a motorbike, be aware that wildlife can be on the same track as you. Snakes can use a cleared path to sunbake, kangaroos and wombats may prefer to stride along the bush track rather than bash their way through the bush, and birds may like to swoop you for invading their territory. Feral animals can also appear almost anywhere.

We mention how to avoid colliding with wildlife when driving a car earlier in this chapter (refer to 'Encountering wildlife on the road'). The same advice applies to when you're riding a motorbike.

When riding a bicycle or motorbike:

- **Stick to the path:** During the day animals can be sleeping in grasses close to the edge of the track.

- **Avoid riding too close to an animal:** Try not to ride over a snake or lizard you encounter on your ride, even if it appears dead. Steer well clear of it if you can, otherwise it may rear up to bite you.

Before you hop on that bike . . .

Planning ahead is paramount. Weather patterns can change with little or no warning, and the remoteness of some areas can make conditions for riding dangerous. We recommend that you don't ride alone and always inform someone of your travel route and times.

Getting around quickly

If your time is limited, catching a plane between capital cities enables you to cover the distance; and the aerial view of Australia's landscape can be fascinating. Travelling during the week is usually cheaper. Shop around for the cheapest flights using the Internet — bargains are available.

If you prefer to stay on the ground, a bus pass or rail pass is a great way to explore Australia — you can take direct routes or travel at your leisure. Buses go just about everywhere and many companies sell passes that allow you to get on and off along the way — so you can spend time at different attractions and still make your major destination. You may even get a glimpse of some of Australia's wildlife.

Some amazing, once-in-a-lifetime rail journeys to experience in Australia are:

The Indian-Pacific: Runs from Sydney on the east coast to Perth on the west coast, via Adelaide. Three days and nights on-board, covering 4,352 kilometres (2,704 miles). This trip includes crossing the Nullabor, which has the longest stretch of straight track in the world — 478 kilometres (297 miles).

The Ghan: Travels 2,979 kilometres (1,851 miles) from Adelaide, South Australia, to Darwin in the Northern Territory, passing through Alice Springs in the middle.

The Overland: Travels between Adelaide, South Australia, and Melbourne, Victoria — 828 kilometres (515 miles).

The bus and rail services are regular and good value, and the vehicles are clean, comfortable and safe.

Bike routes of worldwide repute

Want a bike ride that will emblaze itself on your memory forever? Try one of these trips:

Queensland:

✔ The River Ride Experience: One of the best ways to explore the Brisbane River.

✔ Boondall Wetlands: Located on the northern side of Brisbane, this is a great place to cycle and observe migratory water birds.

New South Wales:

✔ Faulconbridge Point in the Blue Mountains: Breathtaking views over the Blue Mountains are a feature of this scenic route.

✔ Homebush, Sydney Olympic Park: The 8-kilometre Olympic Circuit takes riders past buildings, arenas and areas that were used during the 2000 Olympic Games. In the Overflow, you'll see the cauldron lit by Cathy Freeman in the opening ceremony. The 15-kilometre River Heritage Circuit winds along the Parramatta River.

Victoria:

✔ Yarra Boulevard City Centre to Kew: One of Melbourne's favourite road training routes.

✔ Bay Trail: A seaside ride that will knock your socks off.

Tasmania:

✔ Mount Wellington Descent: A thrilling downhill ride from the peak to the centre of Hobart.

✔ The Tasmanian Trail: Much of this trail is through the state forest between Devonport and Dover.

Western Australia:

✔ Kings Park: This ride offers fantastic views of the city.

✔ Rottnest Island: You can hire a bike, cycle around the whole island and see the *quokkas*, which are small rock wallabies.

Northern Territory:

✔ Fannie Bay: A bicycle and walking track that winds its way through a spectacular mangrove forest.

✔ The Esplanade: A beautiful area with interesting displays of Darwin history splashed along the path.

We include a checklist of clothing and equipment you should consider taking with you in Chapter 3, but always remember to plan ahead; pack maps of where you're travelling to, know where help is available and carry the means to make contact with someone (for example, a charged mobile phone). Never travel alone and always carry water and some food, a bike repair kit and a spare tyre (or two if you're travelling on lesser known tracks).

Setting Up Camp

It doesn't matter whether you're walking, driving or riding — camping is a great way to experience and discover the Australian bush first hand.

One benefit of staying overnight is that you can observe many of Australia's nocturnal animals (wombats, possums, bats, bilbies) and hear the wonderful and unique calls of Australian mammals and bird life: The 'barking' of Koalas, the cackle of kookaburras and yowls of Dingoes.

Camping options

Australia has over 2,800 listed camping grounds and holiday parks. In addition, you can camp in many of the national parks and state forests, and also on Crown land. Some camping sites offer lots of facilities and others don't — you need to decide what suits your needs.

Planning ahead is important if you want the best camping experience. In summer, many popular camping grounds are full. Several national parks and state forests have very limited numbers of camping sites that are 'balloted' off before the season starts. Camping on Crown land is free, but sites can be difficult to locate or are very popular. Tourist information centres are a great source of information on all forms of accommodation.

Crown land is land owned by the government for public use. It includes reserves, state forests and national parks, as well as tracks of land running alongside rivers.

Camping safely

This section gives you a few tips to make that out-of-the-way camping trip a safe and enjoyable experience.

Choose a good camp site

Pick a camp site with some shade — bushes and young trees can act as a wind break and protect you from the sun. Avoid camping under large, older trees, because high winds can easily bring down branches or even uproot the whole tree. Tall trees are also more prone to lightning strikes.

Avoid old river beds, dried-up water courses, narrow canyons and the banks of rivers that can go from no water to a flash flood in a matter of minutes. And avoid camping close to a cliff's edge or on ledges. The edge may not be so easy to see at night!

Don't pitch your tent near stagnant water either, otherwise you'll be plagued by biting insects. And be careful not to put your tent over or near an ant nest.

Ensure a water source is nearby

You need water for all your drinking, cooking and cleaning purposes, and you don't want to have to walk far to get it — water is heavy. On the other hand, don't set up camp too close to the water in case snakes, stock or other creatures that could be a threat to you stop for a drink. A minimum of 50 metres (about 55 yards) from the water's edge is recommended.

Vary your water collection routine and location (some animals, such as crocodiles, will remember your movements and may be waiting to meet you the next time you come back).

Find a suitable spot for cooking

Never cook inside your tent! Food smells can attract goannas, feral pigs, goats, dogs and the like. Store food in containers with lids to avoid visits from animals.

Find a flat area with shade (or use a tarp to make shade) away from overhanging branches that could catch fire. Cooking can be hot work.

And never leave an unattended campfire burning.

Remember your garbage

By keeping your camp site clean, unwelcome animals are less likely to visit. Never leave food or leftovers at the camp site. Wildlife find it irresistible. Collect all litter and burn it or use the lidded rubbish bins provided. Ants, bees, wasps, rats and native mice have a keen sense of smell, so cover or can all opened foods. Snakes and goannas can also be attracted by cooking smells and the presence of mice.

Dealing with pesky animals

Camping should be lots of fun, so getting a good camping site and ensuring it remains pest free is worth the effort. No one wants to spend the whole time warding off insects or watching for snakes and spiders.

Here are some creatures to watch for:

- **Flies, mosquitoes and other flying insects:** You can keep these bugs at bay by burning citronella candles or sprinkling lavender oil on a cloth and leaving it in the open air. To protect your skin, wear a long-sleeved shirt and pants, and find a spray-on or roll-on repellent for your skin type.

- **Ants:** Check the area where you'll be sitting for signs of ants (ant trails, ant mounds and nests). Look up as well as down — ants can be a real pain in the . . . neck.

- **Spiders, scorpions and snakes:** Before getting into bed, check your sleeping bag for creepy-crawlies and snakes, and close your tent up tightly.

Camping out in the bush or outback, listening to the night sounds and smelling the fresh country air is an experience you don't want to miss. But follow the guidelines throughout this book to keep safe, and when packing to return home, make sure you leave the camp site as you found it.

Chapter 3

Preparing for What Nature Serves Up

*A*ustralia is like no other country on earth, offering a treasure-trove of travel experiences and a multitude of amazing creatures to discover. You have so much to explore and enjoy, especially if you're well prepared.

When you take a trip, whether it be an outing to the park, a day trip to the beach or a holiday to another state, careful planning can make or break the experience. If you think ahead and make plans for your journey, you're much more likely to enjoy your time away. Part of this planning involves thinking about the creatures you could encounter. Dabbing on some insect repellent might mean you arrive back home without being covered in bites. Taking and wearing the appropriate clothing might mean you avoid being stung by a jellyfish. Knowing how to behave if you come face to face with a kangaroo, feral pig or snake might mean you come back unharmed and alive.

In this chapter we help you prepare for your trip in advance. We provide advice, guidelines and checklists so that you're all geared up to enjoy the Australian outdoors (and to ensure that creatures don't venture indoors, too).

Planning Your Trip

Hey, time for a holiday, but where do you want to go?

The longer or more adventurous the trip, the more preparation you need to do. You can use this book to help plan a safe trip that meets your wildest expectations, and throughout this chapter we give you resources to ensure you're well prepared. To get started, though, you need to develop your travel plan and think about:

- ✔ What would you like to see or take part in? (Ask yourself, what are your interests?)
- ✔ How do you want to travel?
- ✔ How many people are in your group, what are their ages and does anyone have any special requirements?
- ✔ When do you want to go and what's the weather going to be like?
- ✔ What clothes are you packing?
- ✔ What dangers could you encounter on the trip?

Some of these questions only you can answer. Others are more travel-specific — the type of questions we love to answer.

Letting others know where you're going

It makes a lot of sense to leave these details with trusted family members or friends when you travel:

- ✔ **Personal details:** Your passport number, credit card numbers, student numbers, phone numbers and email address.
- ✔ **Itinerary:** Travel dates and accommodation details, including hotel, motel and youth hostel bookings.

Stay in regular contact with someone at home. Make arrangements to phone, email or send your family or friends a text message once a week. That way, if something goes wrong — say you lose your passport, fall over and get concussion, get lost or get bitten by a snake — someone's already looking for you.

The other thing to do is to let someone more immediate know your present travel plans: Tell the hotel concierge where you're staying, for example, or the person you're presently sharing your room with. Ask them to watch out for your return, especially if you're travelling to a remote place. If you do encounter trouble, the sooner the alarm is raised, the quicker you'll be found.

To ensure people's safety, many popular national parks and reserves ask visitors to sign in and out before heading off on a trail. Make sure you do this, and don't be blasé about it. Give them all the details you can — your life may depend on it.

Watching the weather

Because Australia is a big country, the climate varies immensely from one place to another, as well as from season to season. The weather may be wet and tropical, dry and hot, cold and wet, or even snowing, depending on where you are. You need to pick your dates carefully, plan ahead and check conditions before you travel to ensure you don't get stranded by the weather (and possibly surrounded by active, hungry and dangerous animals).

You can get the information you need about the local weather conditions from tourist information centres or tourist bureaus, local radio stations or online from the Bureau of Meteorology at www.bom.gov.au.

A place of many seasons

Seasons in Australia are the reverse of North America and Europe.

- ✔ **Summer:** December to February
- ✔ **Autumn (or fall):** March to May
- ✔ **Winter:** June to August
- ✔ **Spring:** September to November

But it's not enough to know when the traditional seasons of summer or winter take place. Other seasons affect specific regions in Australia:

- **Wet and dry seasons:** The Top End — the northern parts of Western Australia, the Northern Territory and Queensland — has only two distinct seasons: The *wet season*, from November to April, and the *dry season*, from May to October. During the wet season, crocodiles are breeding and are at their most dangerous, and monsoonal rains can cut off major roads. Also, the deadly Australian Box Jellyfish and other stingers are present in tropical coastal waters during the wet season. Therefore, the best snorkelling and scuba diving season is between May and October, coinciding with the dry season.

  If you visit the tropics in the wet season, avoid swimming in the sea without a full-body lycra stinger suit, but never swim in waters where crocodiles lurk. For more details, see Chapters 4 and 10.

- **High fire danger season:** South of the tropics, a high fire danger season — when bushfires can rage out of control — usually runs from January to February. During this period, some national parks may close, especially on hot and windy days. And during this time, snakes and other reptiles are very active — even at night.

- **Wildflower season:** The season when wildflowers bloom varies from state to state, but brings on numerous honey bees and ants.

- **Ski season:** The ski season in the alpine areas of New South Wales, Victoria and Tasmania runs from late June to September.

Dress sense and the weather

If you're inappropriately dressed and ill-equipped for the weather, you may have a miserable time and even put your life in danger. Be aware of the weather conditions in the area you're travelling to, as well as the type of creatures you may encounter. Even if the weather's really hot, wear long trousers and long-sleeved shirts for bushwalking to prevent leeches or ticks latching onto you, and to provide some protection from snake bite.

Old *bushies* — Australians that live in bush country — sometimes don a unique form of headwear: A wide-brimmed hat with a row of corks suspended from its rim (as shown in this icon). The moving corks keep the flies away, the wide brim shades the face and neck and the bowl of the hat can be used to carry water. It may look a little weird, but it's highly practical.

Watching the time

Keeping a careful eye on the time is always important. No one wants to get caught in unfamiliar terrain when daylight begins to fade. Not only will you find it hard to see where you're going, but also what's out there with you.

Australia spans three time zones. Also, most states switch to daylight-saving time between October and March. Make sure your watch is set to the correct local time, and check the local daily newspaper for sunset and sunrise times.

Getting good maps

Tourist information centres (also called tourist bureaus by locals) are located in all major centres and tourist regions, and are a great source of free maps. Tourist information centres are easy to find: Just look for the blue and yellow sign, as shown in Figure 3-1.

When you hire a car, street directories and maps are usually included. If they're not, ask for them. Motels and hotels usually have free tourist brochures, information and maps on display, so you just help yourself. Often, a whole heap of free booklets and region guides are available for tourists, too. Remember, if you can't see them, ask for them.

Most national parks also offer free maps, mainly to show you where you can and can't go within the park.

You must stick to the marked tracks, not just to preserve the environment, but also for your safety. *Bush bashing* — going off the track — is dangerous; native wildlife may be used to walkers and vehicles on the tracks, but they won't be expecting to be disturbed in their bush retreat! Wild animals are much more likely to attack or defend themselves if you enter their territory or surprise them.

Figure 3-1: Tourist information centre logo.

Organising Your Gear

What you pack for a trip depends on the time of year that you're travelling, what you're doing and where you're spending most of your time.

Clothing

The summer months (December to February) are usually warm to hot in Australia. Lightweight clothing is all that you need during daylight hours, but pack a sweater or jacket for cool nights or in case the weather changes. Always pack a sunhat, sunglasses and sunscreen if you're likely to spend any time outdoors. Insect repellent is also a good idea to ward off biting flies, mosquitoes and other insects.

The winter months (June to August) call for warmer clothing: Sweaters, long pants and a warm jacket for evenings. You need to dress warmer and wear waterproof (and windproof) clothing in alpine areas, where it snows in winter.

Clothing that can be layered is useful if you intend to travel around the country for long periods of time.

You need sturdy shoes to walk, hike, bike, climb over rocks or wade in rock pools. Too many nasty creatures will bite you or suck your blood if you leave your feet and ankles unprotected! A pair of old sneakers or sandshoes is good for walking on rocky shores or through rivers and swamps.

The whole kit and caboodle

What else you pack depends on what sort of activities you've got planned. What you take camping won't be what you want in a classy hotel! All we can advise here is to tailor the contents of your bags to your needs.

Here's a checklist that covers most of the gear you need if you're going it alone — whether to the tropics or to the snow fields. Take from this list what suits your type of holiday:

Clothing:
- [ ] Footwear — boots, shoes, thongs (flip-flops)
- [ ] Hat/cap/beanie/balaclava
- [ ] Gloves
- [ ] Sunglasses
- [ ] Shorts/skirts
- [ ] Long pants/jeans
- [ ] T-shirts/tops/shirts
- [ ] Sweaters/windcheaters
- [ ] Jacket
- [ ] Underwear
- [ ] Socks
- [ ] Thermals
- [ ] Bathers

Personal items:
- [ ] Insect repellent
- [ ] Sunscreen
- [ ] Toiletries
- [ ] Brush/comb
- [ ] Prescription medicines

Travel gear:
- [ ] Suitcase, travel bag(s) or backpack
- [ ] Plastic bags

- [ ] Water bottle
- [ ] Food and drinks
- [ ] Maps
- [ ] First aid kit
- [ ] Mobile phone
- [ ] Mirror (to attract attention in an emergency)
- [ ] Torch
- [ ] Spare batteries
- [ ] Whistle
- [ ] Repair kit
- [ ] Rope/string
- [ ] Compass

Camping gear, optional extras:
- [ ] Books/guide book
- [ ] Camera
- [ ] Binoculars
- [ ] Tent
- [ ] Sleeping bag/sheets/pillow
- [ ] Towel/tea towel
- [ ] Plates/cups/bowls/cutlery
- [ ] Pots and pans
- [ ] Pen knife
- [ ] Needles/thread/safety pins/ scissors

Note that you should also pack your name and emergency details — somewhere that's easy to find.

TIP

Insect repellent and sunscreen

Many personal insect repellent products are on the market; choose one that suits the person in your group with the most sensitive skin. You can also buy citronella candles and oils that ward off insects around a camp setting. Some people place lavender oil-soaked rags in a can set under the picnic table to keep insects away.

Don't forget to protect yourself from sunburn. Sunscreens have SPF (Sun Protection Factor) numbers on the bottles or tubes. Look for a broad spectrum sunscreen — SPF 30 or higher with protection against UVA and UVB rays. The higher the SPF, the longer skin takes to burn.

Packing it all in

This may seem obvious, but don't forget to pack the items you'll need regularly in the most accessible positions: At the top of your suitcase, in the outside pockets of your pack, in the front pannier on your bike or in the glovebox of your car. In an emergency you don't have time to be rummaging for important items like the first aid kit, the torch or the spare batteries.

Your First Aid Kit

A good first aid kit is brightly coloured, waterproof, well organised, properly stocked for your specific needs and handy at all times. Everyone who's likely to open the kit needs to know how to use each item. Even the best first aid kit is of no use if the person with all the knowledge is the one lying unconscious.

The basic first aid kit

First aid kits vary according to their particular use. For example, a first aid kit for a bush walker will contain different items to a first aid kit for a beach-going family.

All basic first aid kits need to contain:

☐ Adhesive dressings (bandaids) in assorted sizes

☐ Triangular bandages

☐ Non-alcoholic wound cleaning wipes

☐ Sterile dressings in assorted sizes

☐ Safety pins

☐ Disposable gloves

Other useful items to consider including are:

☐ Crepe roller bandages

☐ Blister kit

☐ Scissors

☐ Micropore tape

☐ Tweezers

☐ Plastic face shield or pocket face mask

☐ Aluminium rescue blanket

☐ Thermometer

☐ Burn gel

☐ Antacid tablets

☐ Soluble painkillers (aspirin or paracetamol)

☐ Hot and cold gel pack

☐ Sterile eye wash solution

☐ Matches and candle

☐ Notepad and pencil

A hot and cold gel pack consists of a heavy-duty plastic bag filled with gel that can be heated in boiling water and then used to relieve tired muscles, or cooled in a fridge or freezer, or in a cold river or stream, to reduce pain and swelling from a sting or sprain. Hot and cold gel packs are available from pharmacies, sporting goods outlets and stores that sell camping supplies.

If you're carrying medicines for your family, such as aspirin or paracetamol for pain relief, keep these secure and out of reach of children. Aspirin shouldn't be given to children aged under 16 because of the risk of Reye's Syndrome, a rare but potentially fatal inflammation of the brain.

Also, keep personal medications separate from the first aid kit so that children can't access them. However, tell your travelling companions where you keep these medications (for example, an adrenalin EpiPen for allergic reactions, or insulin for diabetes) and ensure they know how to administer them in an emergency.

Storage suggestions

Here's how to look after your first aid kit:

- ✔ Keep the first aid kit in a dry, cool location.
- ✔ Seal bandages in plastic bags, in case creams or liquids leak.
- ✔ Make sure the first aid kit is easily accessible, and that everyone in the family knows where it is.
- ✔ Some items, such as ointments, may have use-by dates. Ensure these are current.
- ✔ If an item from the first aid kit is used, promptly replace it.

Part II
Exploring the Great Southern Land

Glenn Lumsden

*'Good news! According to the manual,
the bite of the python isn't venomous!'*

In this part . . .

*I*n this part, we go overland and explore the creatures that live on this continent. We take a look at the creatures that are obviously dangerous, like crocodiles, with their large teeth and powerful jaws, and different species of Australian snakes, including those that appear harmless but are, in fact, deadly.

We also check out swooping birds and hungry bats, and round up the animals that you may encounter close to cities or in the outback. We even examine some of the creatures that you can pick up when you're least expecting to, as well as those that are hard to ignore — like mosquitoes.

Chapter 4

Crocodile Country

*C*rocodiles are among the most feared creatures on earth. And it's no wonder — their size, their powerful jaws, the ferocity of their attacks and the terrible wounds that their sharp teeth can inflict certainly add to the fear factor. Because of the slow digestive process of the crocodile, human body parts are often recognisable when a 'rogue' crocodile is captured. And the media's pre-occupation in reporting anything gory and sensational certainly attracts people's attention. However, crocodiles kill less people per year than bees and snakes.

Crocodiles belong to the family *Crocodylidae*, along with alligators, caymans and gavials. Crocodiles are the only members of this family found in Australia.

Australia is home to two species of crocodile: The Estuarine (or Saltwater) Crocodile and the Freshwater Crocodile. Figure 4-1 shows where they live.

The Estuarine Crocodile is often called the 'saltie' by locals, and is the larger and, by far, the more dangerous of the two species — see Figure 4-2. There have been no recorded fatal attacks on humans by the Freshwater Crocodile.

Crocodiles are cold-blooded. They rely on the sun and warm tropical temperatures to kick-start their nervous systems (that's why they don't live in moderate or cool climates). Also, even though they spend much of their time in water, crocodiles can't breathe underwater. They have lungs, and breathe air like we do.

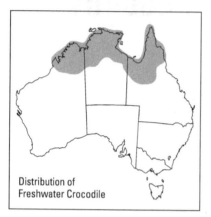
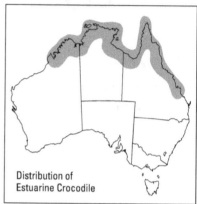

Distribution of
Freshwater Crocodile

Distribution of
Estuarine Crocodile

Figure 4-1: Both species of Australian crocodiles inhabit the northern tropical regions. The Freshwater Crocodile ranges further inland.

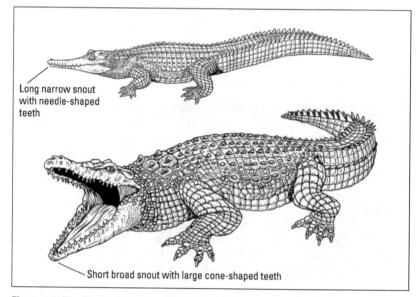

Long narrow snout
with needle-shaped
teeth

Short broad snout with large cone-shaped teeth

Figure 4-2: The Freshwater Crocodile (top) and Estuarine (Saltwater) Crocodile (bottom).

Profiling the 'Saltie'

Estuarine Crocodiles (for a photo, see the colour section in this book) are extremely dangerous and have no fear of humans — in fact, they're one of the few predators that are happy to consider humans as 'food'. Adult male salties average 4 metres (13 feet) in length, but some grow to 7 metres (23 feet) long and weigh more than 1,000 kilograms (2,200 pounds).

As their name suggests, Estuarine Crocodiles are found in estuaries, along beaches and in the open sea. An *estuary* is a wide river mouth into which the sea's tides ebb and flow, mixing saltwater with freshwater. They also live in freshwater and can be found in billabongs, creeks and swamps associated with tidal rivers up to 200 kilometres (124 miles) inland.

A *billabong* is a waterhole or stretch of water that has become isolated from a river or creek during the dry season.

A diet with variety

Salties prey on a variety of animals, including crustaceans, fish, turtles, frogs, birds and small mammals. Big ones will take larger animals, including Dingoes, kangaroos, wallabies, domestic pets and even water buffaloes. Humans are also on the menu — if they're careless enough to make themselves available.

The struggle for survival

When populations of the Saltwater Crocodile are threatened, due to hunting or loss of habitat, recovery is very slow. Only about one quarter of the eggs laid by female salties hatch. The rest don't survive because of flooding, overheating or being eaten by lizards, feral pigs and other predators. Of those that do hatch, most are eaten by other crocodiles, birds or fish. This leaves the odds of an egg producing an adult crocodile at about one in a hundred.

The crocodile roll

Salties don't chew their prey. They crush it with their powerful jaws and swallow it. If their prey is too large to simply crush and swallow, they break it up into smaller chunks by shaking their heads violently or rolling several times in the water.

The Freshwater Crocodile

Although the Freshwater Crocodile (also called the Johnston's Crocodile or the 'freshie') is unlikely to attack humans, still consider this crocodile dangerous — simply because of its size and weight. Freshies average about 1.5 metres (almost 5 feet) long, but can grow up to 3 metres (10 feet) in length. If a large Freshwater Crocodile is provoked or cornered, it can bite or cause serious injuries by thrashing its tail around. The rare attacks on humans that have been recorded by the Freshwater Crocodile are believed to be the result of provocation or mistaken identity.

Freshwater Crocodiles usually live in freshwater billabongs, rivers and swamps. But they can also tolerate the saltier water of estuaries. They can easily be distinguished from the more dangerous Estuarine Crocodile by their smaller size, narrower snout and needle-shaped teeth.

The Freshwater Crocodile feeds mainly on fish, frogs, lizards and turtles, but a larger one will eat small mammals, including wallabies and birds. The Freshwater Crocodile is less territorial that the Estuarine Crocodile and can congregate in large numbers, especially during the dry season when they're breeding.

Heed the Warnings

If you're living or holidaying in the tropical north of Australia, you can't miss the numerous signs warning you not to swim in the sea, rivers, creeks and waterholes where crocodiles are likely to be present. *Don't ignore the warnings!* Even if no signs are visible you're best to stay out of the sea and any river, creek or waterhole within about 200 kilometres

(125 miles) of the tropical north coast. You can never tell whether a crocodile may be lurking — especially at night.

Surprise attack

Salties, especially, may look slow and clumsy, but they can move with lightning-fast speed over a short distance of up to a couple of metres. They quietly observe their prey and wait until it's close enough for them to launch a sudden and powerful attack. When attacking from the water, a saltie uses its muscular tail to propel itself swiftly through the water. It swims with only its eyes and nostrils above the water, so the victim is unlikely to see the attack coming.

You're not safe in dark murky waters at night-time, either. Salties have excellent night vision (even under the water), an acute sense of smell and can detect vibrations through the water.

Defending their territory

Estuarine Crocodiles are territorial. In areas where food is plentiful they live in communities led by a dominant male. A typical community might include two mature breeding females and several immature crocs, and sometimes some young sexually mature males who would be kept away from the breeding females by the dominant male. Eventually the younger males are forced out by the dominant male to find their own territories — or one may replace the dominant male.

Basking in the sun

During the day, Estuarine Crocodiles often bask in the sun on the banks of rivers, creeks and waterholes with their mouths open wide. This enables the crocodile to maintain a body temperature of between 30° and 32° Celsius (between 86° and 90° Fahrenheit). They lie so that they can absorb just the right amount of the energy from the sun. If they get too hot they open their mouths to allow air to circulate inside, move into the shade or slip into the water.

The Croc Hunter: Steve Irwin (1962–2006)

Steve Irwin brought a completely new meaning to the term 'crocodile hunter'. Before the 1940s, crocodile hunters came close to completely wiping out the populations of crocodiles in the wild. Now, a new breed of crocodile hunter in Australia is helping to ensure that these giant creatures with a link to the age of dinosaurs survive in the wild.

Steve Irwin started catching crocodiles with his father Bob at the tender age of 9. Many of these crocs were taken to his father's Beerwah Reptile Park on the Sunshine Coast of Queensland, safely away from the hunters who wanted their skins. Beerwah Reptile Park has since evolved into Australia Zoo — it now cares for a diverse range of animals from Australia and all over the world.

Before he took over running Australia Zoo from his father, Steve worked as a volunteer, catching and relocating large crocodiles that were too close to human populations, to more remote areas or to the zoo.

In 2002, Steve and his wife Terri established the charity, Wildlife Warriors, to support the protection of injured, threatened or endangered wildlife. This group gives anyone interested in the conservation of wildlife the opportunity to support their cause.

You can find out more about Australia Zoo at www.australiazoo.com.au and Wildlife Warriors at www.wildlifewarriors.org.au.

Female crocodiles lay between 40 and 60 eggs in a nesting mound of vegetation, soil and mud above the water's edge. Salties — male and female — vigorously defend their territories and their nests against intruders, including other crocodiles and humans. They're at their most dangerous — especially the dominant male — during breeding season (September to April), which mostly coincides with the wet season.

Minimising the risk of crocodile attacks

Crocodile attacks on humans, although rare (on average about one per year in Australia), usually result in serious injuries or death. If you're in crocodile country you can avoid an attack by taking these steps:

- Heed the warning signs and don't ignore verbal warnings from locals.

- Swim only in areas that are recommended by local authorities for swimming.

- Never provoke or disturb crocodiles, regardless of size. And don't even think about looking for or interfering with their eggs.

- Never feed crocodiles. This is not only dangerous, but also illegal.

- When fishing, stay a few metres away from the water's edge at all times. Clean fish in a bucket at least 50 metres (about 55 yards) from the water's edge and away from camp sites or boat ramps.

- Set up camp at least 50 metres (about 55 yards) from the water's edge and further away from places where wildlife or stock drink. Never leave any food scraps or bait at your camp site.

- Don't prepare food or wash dishes near the water's edge. It's safer to do this at your camp site. If you must collect water from a river, creek or waterhole, do so quickly from a different place each time. Salties may be watching you.

- Stay away from any marks left by a crocodile sliding down a bank. The crocodile that made the marks may return.

- When you're in a boat, keep your arms or legs well inside. If you fall overboard, get out of the water as quickly as you can.

Meet Mr Freshy

The only Freshwater Crocodile at Australia Zoo, on the Sunshine Coast in Queensland, is Mr Freshy, believed to be about 130 years old.

Mr Freshy was rescued by Steve Irwin and his father Bob in 1970 after he was shot by hunters, losing his right eye.

Most crocodile attacks on humans occur when people ignore warning signs, or swim in rivers or creeks where salties are likely to be lurking. In some cases, attacks are the result of foolish behaviour by humans. A classic example of this occurred in 2006 when a group of tourists ignored warning signs and waded into a creek while a large saltie was basking on the bank. One tourist tried to attract the attention of the crocodile for a photograph by slapping the water with a stick. He certainly got the crocodile's attention and was lucky to escape with only teeth marks on his leg. The unfortunate result of this attack was that the crocodile was removed from his natural habitat and relocated to a crocodile farm.

If a crocodile attack takes place in the water, get the victim out of the water. Apply pressure with a rolled-up towel to the wound to try and stop the bleeding. Raise any bleeding limb; reducing blood loss is critical. Keep the victim as calm and warm as possible. Call for an ambulance immediately. Avoid moving the patient while waiting for the ambulance — movement increases blood loss.

Protecting Crocodiles from People

Both the Freshwater and Estuarine Crocodile are *protected species* in Australia. This means that it's illegal to interfere with them, their nests or their eggs in any way without a permit. Both species were hunted to the brink of extinction for their skins until 1942 (in the case of the Estuarine Crocodile) and 1963 (in the case of the Freshwater Crocodile).

Although their populations have recovered, both species face a new threat: Loss of habitat due to the growth of human populations in the tropical north of Australia — especially along the east coast of Queensland.

As human population and tourism in crocodile country grow, encounters between humans and the deadly Estuarine Crocodile are more likely. So, being respectful of salties and their habitats, and knowing how to avoid an attack, is increasingly important.

When dangerous Estuarine Crocodiles are discovered in populated areas, and are likely to be a threat to people, they're captured and relocated to more remote areas — sometimes to crocodile farms or zoos.

Crocodile farms

Although crocodiles in the wild are protected from being killed by humans for their skins, the demand for crocodile skin and meat still exists. Estuarine Crocodiles are now bred in farms for this purpose. Crocodile farms were initially stocked by crocodiles from the wild, including some large ones that were a threat to human populations. However, crocodile farms are now largely self-contained and provide their own breeding stock. Some of these farms are tourist attractions and play a role in educating the public about crocodiles. For example, Hartley's Creek Crocodile Farm, between Cairns and Port Douglas, Queensland, runs educational crocodile farm tours, along with crocodile feeding and attack shows.

Chapter 5

Snakes: Legless and Lethal

In This Chapter

▶ Revealing the deadly side of snakes

▶ Understanding identification problems

▶ Avoiding dangerous snakes

▶ Staying calm during a close encounter

▶ Unmasking Australia's most venomous snakes

*S*nakes are responsible for more human fatalities in Australia than any other animal. Between 1979 and April 2007 there were 67 recorded snake-bite fatalities. This is more than those caused by crocodiles and sharks combined.

Like all reptiles, snakes have scales and a body temperature that changes with their surroundings. Other reptiles include lizards, tortoises, turtles and crocodiles.

If you encounter any snake, consider it dangerous. As we explain throughout this chapter, don't assume that you can accurately identify it — even the experts find identifying snakes difficult sometimes. So even if you think you're looking at one of the harmless varieties, there's a good chance you're wrong. (To find out why, check out 'The Problem of Identification' later in this chapter.)

We describe the most dangerous Australian snakes in this chapter, but there are many others that have not been known to cause illness, injuries or death in the past, but might have the potential to be lethal. For example, the Rough-scaled Snake, covered later in this chapter, was considered harmless until the 1960s, but its venom has since been identified as toxic.

Almost all Australian snakes live and hunt on the ground. But two families of snakes live in the sea. To find out more about sea snakes, see Chapter 11.

Snakes Alive and Deadly

Snakes are *carnivores* — that is, they eat other animals. Their diet usually consists of frogs, mice, rats, lizards, birds and other snakes. Each species uses one of two methods to kill or disable its prey.

- ✔ **Venom:** A mixture of toxins that can paralyse, interfere with blood circulation or damage muscles and other tissue.

- ✔ **Constriction:** This involves squeezing the prey, causing it to suffocate. The most well-known constrictors are the Python group and Boa Constrictors, which are native to Central and South America. The keeping of Boa Constrictors as pets is strictly prohibited in Australia.

Obviously, people are not suitable prey for snakes, but if you provoke them, either deliberately or accidentally, they may attack — by biting and injecting venom through their fangs or by constriction. Fortunately, most snakes are not venomous or powerful enough to kill humans, but the dangerous minority can be lethal. Australian constrictors are not powerful enough to kill humans — except young children and babies.

Although Australian constrictors are often described as harmless, they're capable of killing small animals. Pythons look dangerous because of their length — typically they grow to about 3 metres (10 feet) long, but one common species, the Scrub Python, can grow up to 7 metres (23 feet) long. Pythons are notorious for killing and eating chickens and have been known to kill pets, like cats and small dogs. They grasp them with their teeth (constrictors don't have fangs) and literally squeeze them to death, before swallowing them whole.

The toxins that make up snake venom vary from species to species. Venom can include one or more of the following toxins:

- ✔ **Neurotoxins:** Attack the nervous system, causing vision problems, difficulty swallowing and speaking, numbness, spasms and paralysis. Paralysis often leads to respiratory failure.

- ✔ **Myotoxins:** Damage muscle tissue, sometimes leading to kidney or heart failure.

- ✔ **Pro-coagulants:** Strip the blood of clotting factors, which increases a tendency to bleed more easily. The most dangerous consequence is an increased risk of bleeding in the brain.

- ✔ **Anti-coagulants:** Cause blood thinning, increasing the risk of bleeding.

- ✔ **Necrotoxins:** Cause blistering and tissue damage near the site of the bite.

Fortunately, scientists have developed antivenoms to counter the effects of all Australian snake venoms. Only about 10 per cent of snake bite victims need to be treated with antivenom because the amount of venom injected by a snake is usually small. In many cases the bites are 'blank' and no venom is injected.

Antivenoms can be injected into veins or into muscles to treat poisoning caused by the bite of a venomous creature. Antivenoms are available for most of Australia's venomous animals.

Protecting snakes from people

All snakes — even the deadliest ones — are protected by Australian law. Killing them or removing them from their natural habitat is illegal. A permit is required to keep any snake in captivity.

Frightening fangs

A snake's venom is injected through its fangs (see Figure 5-1), which are so sharp and thin that fang marks can be painless and hardly visible on the victim. Many victims don't even know they've been bitten until serious symptoms occur. Some snakes have significantly longer fangs than others, increasing the risk of a serious bite. The amount of venom injected can be controlled by the snake, using muscles to control the pressure on the venom gland, which produces the venom. A venomous snake has one venom gland on each side of its head.

Snakes also have teeth that point inwards. Their major purpose is to keep prey moving in the right direction as it's being swallowed.

Home s-s-s-sweet home

In general, snakes are likely to be found wherever they can find food and warmth. You can find them basking in the sun on cool days, but they don't like really hot conditions. They find shady places to escape the heat of the sun when it gets too hot. But each species has its own preferred habitat. Some species are only found in a very small region — for example, the Broad-headed Snake is common only in Sydney and surrounding areas. Others, such as the Mulga Snake and Common Death Adder, can be found throughout most of Australia. (We talk more about individual species in 'Introducing Australia's Most Dangerous Snakes' later in this chapter.)

Swallowing it whole

Snakes can swallow prey much larger than their own heads. Mice, rats, possums and even wallabies can be swallowed whole. The lower jaw bones of a snake are only loosely attached to its skull and each bone can move independently. This allows the snake's mouth to stretch over the prey until it's swallowed.

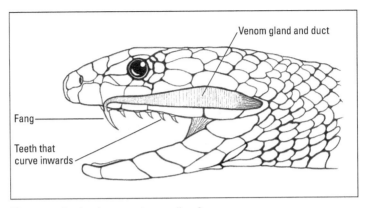

Figure 5-1: Snakes have teeth as well as fangs.

Snakes occasionally venture into urban areas and farms in search of food and shelter, so you need to be careful everywhere, not just out in the bush. (See Chapter 15 for more information on unwelcome gatecrashers.)

The Problem of Identification

If you encounter a snake in the bush, identifying it can be difficult. Even snake experts find some species hard to identify. The problem is that the most obvious characteristics of a snake — its length and colour — are not accurate identifiers.

Sizing snakes up

Size is of little help when identifying a snake — nor is it an indicator of how deadly the snake is. For example, the adult Black Tiger Snake can range from 90 centimetres (3 feet) up to 2.4 metres (8 feet) in length. The deadliest Australian snake, the Eastern Brown Snake, is usually less than 1.5 metres (5 feet) long but can be as long as 2.4 metres (8 feet). Australia's longest venomous snake is the Coastal Taipan, with an average length of 2.5 metres (a little more than 8 feet).

Changing colour

Colour varies considerably within most species of snakes. And the colour of an individual snake can change as it gets older — stripes and specks can completely disappear. To make matters worse, when snakes shed their skin, as they do several times each year, the newly revealed skin is brighter than the discarded skin. The world's most venomous snake, the Inland Taipan, even changes colour with the temperature of the air and ground.

What's in a name?

Even snake names can be misleading. For example, the Mulga Snake is also known as the King Brown Snake. But its colour ranges from brown to black. It's classified by zoologists as a black snake and the antivenom used to treat its bite is black snake antivenom. The Black Tiger Snake has no stripes, but is black all over. And the Red-bellied Black Snake can actually have a cream belly.

Avoiding Unexpected Encounters

The difficulty of identifying snakes becomes most serious if you encounter one unexpectedly — or worse still, get bitten. You're not likely to know whether it's deadly, slightly dangerous or completely harmless. Nor will you know if the snake is more likely to flee or attack you fiercely. So you're best to treat all snakes as potentially dangerous.

Many bites are the result of accidentally standing on or getting too close to a snake. Such accidents can be avoided if you follow these guidelines:

- **Watch where you walk.** Use cleared tracks and paths wherever possible and take care while stepping over fallen logs.
- **Use a torch while walking at night.** Some snakes — including some of the deadliest ones — are active on warm nights as well as during the day.

- ✔ **Wear sturdy shoes or boots and long trousers.** Thongs (flip-flops), sandals and shorts are not appropriate for walking in the bush. Long trousers reduce the risk of a direct bite on the leg.
- ✔ **Tread heavily in the bush.** Although snakes can't hear, they feel vibrations in the ground. So, if you tread heavily, in solid footwear, snakes in your path are more likely to slither away before you get there.
- ✔ **Avoid walking in long grass.** Don't run through long grass, either. Snakes 'hang out' in long grass to protect themselves from predators.

Some bites occur when the victim accidentally touches a concealed snake. Avoid this by:

- ✔ Keeping your hands out of hollow logs, cavities in rocks, ground cover or long grass.
- ✔ Checking your sleeping bag and clothes before getting into them when you're camping. Also check towels, rugs and blankets before use.

Surviving an Unexpected Encounter

If you do accidentally stumble across a snake, the solution is simple.

- ✔ Assume that the snake is dangerous and leave it alone. Never try to catch it or kill it. If you do, you're likely to get bitten.
- ✔ Stay completely still. If the snake doesn't slither off, back away, very, very slowly.

Don't be fooled into thinking that a dead snake is safe to pick up. People have been bitten by dead snakes. The biting reflex is still there for hours after they die. And even after the biting reflex fades, the venom is still toxic. If you scratch yourself on the fang of a long-dead snake — even a stuffed one — you're still in danger.

If you or a companion is unlucky enough to get bitten, regardless of what type of snake you think it is, follow these guidelines:

- ✔ Don't wash the area around the bite — the venom left on the skin can help doctors select the correct antivenom.
- ✔ Keep the victim as calm as possible.
- ✔ Apply a pressure immobilisation bandage immediately (see Appendix A) even if the victim is displaying no visible symptoms.
- ✔ Seek medical attention urgently.
- ✔ Don't cut the area around the bite to make it bleed.
- ✔ Don't try to suck out the venom.

Introducing Australia's Most Dangerous Snakes

Australia is home to more than 100 species of snakes, of which about 25 are considered to be dangerous. Here are the most deadly Australian snakes, according to the number of deaths they've caused:

1. **Eastern Brown Snake**
2. **Eastern (or Mainland) Tiger Snake**
3. **Coastal Taipan**

None of these three snakes has the most toxic venom. The most venomous Australian snake is the Inland Taipan. The danger of a snake species to humans doesn't just depend on the toxicity of its venom. It also depends on

- ✔ The amount of venom the snake injects.
- ✔ How close the snake's habitat is to human populations.
- ✔ Whether the snake instinctively flees or attacks a threat.

Beware of the brown

Brown snakes can be found throughout mainland Australia. As their name suggests, they're usually brown, but some can be black, olive or orange in colour. Some of them have dark spots or blotches. They tend to be more lightly coloured underneath. Brown snakes are most active during summer and are rarely active at night. Brown snakes wind themselves up into a distinctive 'S-shaped' defensive stance if they're threatened.

The deadliest of the brown snakes, the Eastern Brown Snake (see photo in colour section), is also known as the Common Brown Snake. This snake can be found in eastern and central Australia. Adults have an average length of about 1.5 metres (about 5 feet). The Eastern Brown Snake is very fast and unpredictable — you can never tell whether it will attack or flee when encountered. And if it does attack, it could bite and slither away or bite and hold on. Victims of brown snake attacks often don't even realise they've been bitten. Its venom contains a very toxic neurotoxin that causes paralysis and a pro-coagulant toxin that increases the risk of bleeding, which can lead to sudden and unexpected death.

Here are some other potentially deadly brown snakes:

- ✔ **Spotted Brown Snake:** This snake, also known as the Dugite, is found in the southwest region of Western Australia and along the western half of the south coast. It hisses loudly when provoked and strikes repeatedly.

- ✔ **Western Brown Snake:** The Western Brown is also known as the Gwardar or Collared Brown Snake, and is found throughout Australia, except for the southeast and southern coastal regions.

- ✔ **Ingram's Brown Snake:** Although confined to the Barkly Tablelands region of the Northern Territory, this snake also lives in parts of inland northern Queensland and the northeast corner of Western Australia.

Not quite so deadly, but still dangerous are the very small Five-ringed Brown Snake (western and central Australia), the very shy Speckled Brown Snake (inland Northern Territory,

Queensland and South Australia) and the dark brown
Peninsula Brown Snake (the Eyre Peninsula of South
Australia).

Tiger snakes — stripes are optional

Tiger snakes are so named because they usually have light
bands that look like a tiger's stripes around their bodies. The
bands can be yellow, light brown or grey. But in some cases
no visible bands exist. Of course, this makes identification
very confusing.

If you're bitten by a snake and need antivenom, a description
of the snake is not as useful as the venom left on your skin.
Length and colour are not reliable identifiers of many snakes.

Although tiger snakes are only found in relatively small
regions of Australia (see Figure 5-2), they're the snakes city-
dwellers are most likely to encounter. They feed primarily on
frogs and spend a lot of their time near rivers, creeks, lakes
and dams. Their distribution also coincides with the most
heavily populated areas of Australia and they're not too shy
about visiting farms and outer suburban homes in search of
mice and rats.

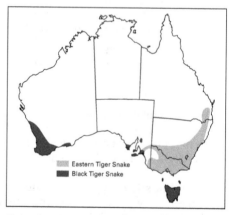

Figure 5-2: Tiger snakes are confined to relatively small, heavily populated
regions of Australia.

Tiger snake venom is highly neurotoxic, causing paralysis that sometimes leads to respiratory failure and death. Its effects may become apparent within a few minutes of the bite, beginning with a feeling of nausea accompanied by dizziness, headache and vomiting. The venom also causes muscle damage, which can lead to kidney failure and death. Tiger snake venom also contains pro-coagulants that increase the risk of bleeding.

The Eastern Tiger Snake (see photo in colour section), also known as the Mainland Tiger Snake, is found in Victoria, eastern New South Wales and a tiny area of southeast Queensland. Adults average about 1 metre (about 3 feet) in length. These highly venomous snakes have broad heads and can be any colour from light brown to black. The tiger-like bands can be various shades of yellow or completely absent.

The only other species of tiger snake in Australia is the Black Tiger Snake. Its venom is not quite as toxic as the Eastern Tiger Snake but can still be lethal to humans. As you can guess from its name, the Black Tiger Snake is black. It's usually grey underneath and some have yellowish bands. Several subspecies of the Black Tiger Snake exist, each living in different regions, and all have alternative names:

- Western Tiger Snake (southwest region of Western Australia)
- Tasmanian Tiger Snake (Tasmania and surrounding islands)
- Chappell Island Tiger Snake (Chappell Islands in Bass Strait)
- Peninsula Tiger Snake (Eyre Peninsula in South Australia)
- Krefft's Tiger Snake (Flinders Ranges in South Australia)

Toxic taipans

Australian taipans are among the most feared snakes in the world. And with good reason. The two most common species — the Inland Taipan and the Coastal Taipan — rank one and three respectively as the snakes with the most toxic venom in the world (see Figure 5-3 for a map of their ranges). Their venom is highly neurotoxic, causing paralysis.

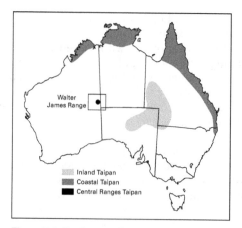

Figure 5-3: The known distribution of taipans in Australia.

Taipans are the only Australian snakes that feed exclusively on mammals — no frogs, lizards or birds for these large and fast-moving reptiles.

Fierce when disturbed

Although the Inland Taipan is also known as the Fierce Snake, this creature is quite shy, but when provoked it attacks with extreme ferocity.

Its colour varies from light brown in summer to a darker brown in winter, sometimes with dark flecks, so it can be mistaken for the Eastern Brown Snake. The Inland Taipan is best known for having the most toxic snake venom in the world, but it has short fangs and injects only a small amount of that venom. Because this snake lives in arid regions, away from most populated areas, it's rarely seen. So, although it's the most venomous of all, it's not the deadliest. There are no recorded fatal attacks; most bite victims are snake collectors.

Shy but deadly

The deadliest taipan is the Coastal Taipan (see photo in colour section), also known as the Common Taipan, which can be found on and near the eastern coast from northern New South Wales and across the north coast as far west as the Kimberley region of Western Australia. It has also been spotted on Fraser Island off the coast of Queensland. The Coastal Taipan is the longest venomous snake in Australia with an average adult length of 2 metres (about 7 feet). The longest specimen recorded measured 3.3 metres (about 11 feet) from head to tail.

The Coastal Taipan likes to feed on rats and mice, so can often be found searching for prey near farm buildings, and in rubbish tips and sugarcane fields. So, take special care in these areas. If you accidentally corner or step on one, the snake is likely to instinctively raise its body in a coil, wave its tail backwards and forwards, and strike viciously.

Although the Coastal Taipan's venom isn't as toxic as its inland relatives, it has longer fangs (up to 13 millimetres or 0.5 inches) long, is capable of multiple bites and injects a huge amount of venom — enough to kill a child within minutes. The Coastal Taipan has caused numerous human fatalities. Before an antivenom became available in 1955 very few people survived its bite.

New kid on the block

It may be hard to believe, but scientists are still discovering new species of snakes. Well, they're not really new; they've just never been seen or identified before. In 2006, a group of scientists surveying animals and plants near Uluru in central Australia found what they thought may have been a Western Brown Snake, and they took it back to the Western Australian museum to examine it more closely. The snake turned out to be a previously undiscovered species of taipan snake and was given the name Central Ranges Taipan.

It adders up to death

Death adders look and behave differently from other Australian snakes. They're heavily built with a distinctive triangular-shaped head and a finely tapered tail (see Figure 5-4). Death adders are relatively short, with an average length of 65 centimetres (about 2 feet), but some have been known to grow to over 2 metres (about 6.5 feet) long.

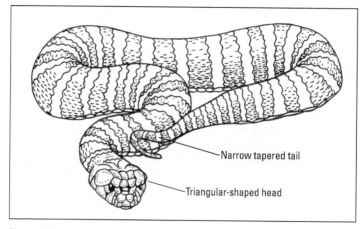

Narrow tapered tail

Triangular-shaped head

Figure 5-4: Death adders are more heavily built than other Australian snakes and have a distinctive triangular-shaped head.

Their colour varies from reddish brown to a dark grey, with darker or lighter and bands. This colouring provides an effective camouflage against the ground while they're resting or hunting. Death adders wriggle their tail to lure their prey before striking quickly and decisively.

Staying put

Unlike most dangerous snakes, death adders don't slither away when humans approach. They remain still, concealed in fallen leaves, sand or gravel, grass and other ground cover. This makes it very easy to accidentally stand on one. If you do tread on one, the snake will flatten its body and strike rapidly and accurately — injecting a strongly neurotoxic venom with its long fangs. Without antivenom treatment, a death adder bite can cause paralysis resulting in death.

Four of a kind (at least)

There are four major species of death adder in Australia:

- ✔ Common Death Adder
- ✔ Northern Death Adder
- ✔ Desert Death Adder
- ✔ Pilbara Death Adder

The Common Death Adder is more heavily built and longer than the other three listed, but is no more dangerous. The smallest of the death adders, the Pilbara Death Adder, rarely measures more than 50 centimetres (about 1.5 feet) long. Figure 5-5 shows where these four death adders live.

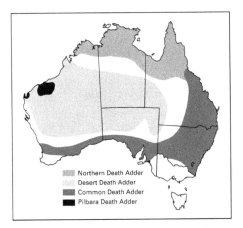

Figure 5-5: Where the four species of death adder live in Australia.

All death adders are dangerous. The best way to avoid being bitten is to wear sturdy shoes or boots while walking in areas where death adders might be resting or waiting for prey. And always use a torch while walking in unlit or poorly lit areas at night.

Black is beautiful — but keep your distance

The bodies of black snakes display colours including brown, grey, red, cream, pink, purple and blue. Their bodies can also be adorned with spots, bands and blotches of a variety of colours. The most dangerous species of black snake averages about 1.5 metres (5 feet) in length.

Although not as dangerous as brown snakes, tiger snakes, taipans and death adders, black snake venom still packs a punch and has been responsible for human fatalities.

The Cane Toad menace

The number of Red-bellied Black Snakes in Queensland and northern New South Wales is decreasing at an alarming rate as a result of eating Cane Toads. The Northern Death Adder is also under threat for the same reason. Cane Toad venom is lethal to all snakes except one — the harmless Keelback (or Freshwater) Snake.

The Mulga Snake — black or brown?

The most widespread of the dangerous black snakes is the Mulga Snake, which is also known as the King Brown Snake. It can be found throughout most of mainland Australia (see Figure 5-6). Mulga Snakes shelter in abandoned animal burrows, and under logs or rocks.

The Mulga Snake isn't normally aggressive but hisses loudly and attacks viciously if it feels threatened. It flattens itself out, arches the front of its body and strikes quickly, often more than once, and sometimes even chews on its victim, injecting more venom than any other Australian snake (yuk!). The venom causes swelling and pain at the site of the bite and extensive and potentially lethal muscle damage.

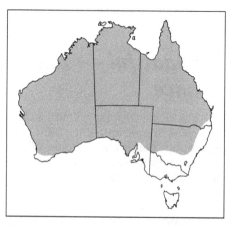

Figure 5-6: The Mulga Snake is found throughout most of the Australian mainland. Other black snake species are confined to smaller regions.

It also contains neurotoxins. Mulga Snake venom has been known to kill humans.

The Mulga Snake is sometimes wrongly identified as a brown snake. Its body colour can be anything from light brown to dark brown — but its other characteristics, including its defensive stance and its venom, match those of the black snake species.

Colourful creepers

Some of Australia's most colourful snakes are black snakes. Their names provide clues about their appearance. Like the Mulga Snake, their venom is toxic and causes swelling and pain at the site of the bite, along with muscle damage. These are the most dangerous:

- **The Blue-bellied Black Snake:** Also known as the Spotted Black Snake, this creature is the most venomous of the black snakes. It is grey to black in colour and has a bluish-grey belly. Most have lighter coloured spots or blotches. The Blue-bellied Black Snake inhabits southeast Queensland and parts of eastern New South Wales. Like the Mulga Snake, it hisses and flattens its body when provoked. It's not known to have caused any human fatalities — most probably because of its limited distribution and because it only injects a small amount of venom.

- **The Red-bellied Black Snake:** Also called the Common Black Snake, this snake is found in eastern Australia from the tropical north all the way down to Victoria. There are also populations in the southeast region of South Australia. As its name suggests, this snake has a black body (with a tinge of purple) and a red belly. It usually has a light grey patch on the front of its head, too. But, as with so many snakes, variations exist. In fact, some Red-bellied Black Snakes don't even have red bellies — they could be pink, cream or even black.

  The Red-bellied Black Snake lives mostly on a diet of frogs, but also enjoys small fish and eels — it's a very good swimmer and can easily be mistaken for a floating stick in rivers and creeks. If its normal prey is not available it is quite happy to make a meal of other snakes — even its own species. As a result, it's not kept with other snakes in captivity.

Even though it's not as widely distributed as the Mulga Snake, the Red-bellied Black Snake is more likely to be encountered by humans because most of Australia's major cities are within the region it inhabits. It tends to flee if humans approach. When cornered it hisses loudly, flattens itself and arches the front of its body. But the snake is usually all bluff — it rarely bites. However, its venom is toxic enough to threaten the lives of small children.

✔ **The Yellow-bellied Black Snake:** Also known as the Butler's Snake and the Spotted Mulga Black Snake, this snake is confined to a remote region in the inland southwest of Western Australia, so it's rarely sighted by humans. It has a glossy black body with a yellow belly.

✔ **Collett's Snake:** This species of black snake is found in remote central Queensland, where it's most active during the wet season. During the dry season it shelters in cracks in the dry soil. Collett's Snake is brown or black with markings, blotches or bands of various colours, including a spectacular crimson, pink, orange or cream. Because it has such a limited range and lives in sparsely populated areas, bites are very rare. Most bite victims have been snake collectors.

When one snake can have up to three different names, no wonder it gets confusing trying to identify them — or at worse, to figure out which species of snake bit you!

Looks are so deceiving

Two less well-known, but very dangerous, snakes are very difficult to identify because they look similar to other snakes. Here are the rogues:

✔ **The Rough-scaled Snake:** This snake, also misleadingly known as the Clarence River Tiger Snake, is found along much of the east coast of Queensland and in northeast New South Wales, usually in or near forests, creeks and swamps. Greenish-brown with dark bands or blotches, at first glance the Rough-scaled Snake sometimes looks like a small tiger snake. But what makes it more dangerous is that it's often mistaken for the harmless Keelback Snake, which inhabits the same regions.

Its venom is very toxic, causing paralysis, muscle damage and bleeding. The Rough-scaled Snake is believed to be responsible for several human fatalities.

✔ **The Small-eyed Snake:** Also called the Eastern Small-eyed Snake, this creature inhabits wooded areas in an approximately 200-kilometre-wide (125 mile) strip along the east coast of Australia. It usually shelters under rocks, logs, leaf litter or sheets of bark. With a shiny black body, which is pink underneath, it's easily mistaken for the less dangerous and more common Red-bellied Black Snake. But its average length is around 50 centimetres (about 20 inches), less than half the average length of the Red-bellied Black. The toxicity of the Small-eyed Snake is variable and causes muscle damage. There has been one recorded fatality.

Copperheads are cool

Copperheads are found in places that are too cool for all other venomous Australian snakes. They're limited to northern Tasmania, Victoria and the highlands of New South Wales. The smallest of the three species of copperhead, the Pygmy Copperhead, is also found in parts of South Australia — mainly in the Mount Lofty Ranges and on Kangaroo Island.

The copperhead's body is copper to black in colour, with a head that sometimes seems too small for its neck. Its belly can be cream, yellow or orange. Its preference for a diet of frogs means that it's usually found near water. Copperheads are good swimmers and can be easily mistaken for floating sticks. As well as frogs, they eat small lizards and other snakes — including their own species.

Because of their tolerance of cool weather, you could encounter copperheads at times when other snakes are hibernating. They are shy snakes and will only bite if cornered or provoked. Their venom is toxic enough to kill a human, but because they strike slowly and their fangs are small, they either don't succeed in biting or can't inject enough venom to knock you off. Their bites are not known to have caused human fatalities.

Chapter 6

Don't Let Size Fool You

*1*f you've ever been a backpacker, or had to stay at some seedy hotel, you already know that some of the smallest creatures can cause the greatest irritation. Australia has its fair share of little critters that bite.

Scorpions sting you because they see you as a threat, but leeches see you as a meal. To ticks, mosquitoes and lice, you are a vital part of their lifecycle: Without a blood meal they can't reproduce and will die.

Although these creatures and some other creepy critters we cover in this chapter, such as centipedes and millipedes, are small, don't let their size fool you. All are capable of being much more than an irritation. Some are capable of making you very ill.

Stinging Scorpions

All scorpions are venomous; their stings kill thousands of people every year in other countries, but only one fatality has been caused by an Australian species — and that was possibly due to an allergic reaction. Scorpion stings can be extremely painful and leave you feeling ill for several days.

Scorpions are members of the *arachnid* class of animals. Other arachnids include spiders, ticks and mites. Scorpions can grow to 12 centimetres (4.5 inches) long, making them some of the largest arachnids found in Australia. Scorpions are easily distinguished from other arachnids by their front pincers, which they use for grasping prey and for defence. They also possess a stinging tail containing venom. When a scorpion is threatened or about to attack its prey, its tail is curled forward, over its body, ready to strike at any time (see Figure 6-1). Australian scorpions are generally harmless and timid, using their sting for killing prey or defending themselves.

A few Australian scorpion species reproduce *asexually,* which means unfertilised eggs develop into new individuals. Most need a mate, though, and engage in a fascinating mating dance that can last for hours (see related sidebar). Female scorpions give birth to live young, which look exactly like miniature adults. The female then carries the young around on her back (as shown in Figure 6-1), until they're old enough to fend for themselves.

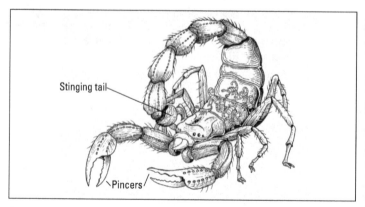

Figure 6-1: A female scorpion carrying young on her back.

The dance of the mating scorpions

Male and female scorpions perform a rhythmic mating dance, somewhat reminiscent of a tango. They recognise each other using *pheromones* (animal body odours) and vibrations.

The courtship begins with the male grasping the female's pincers. The pair then scurry one way then the other, back and forth in a 'dance routine', their bodies juddering and their mouth pieces coming together for a 'romantic' kiss until the male locates a suitable place to deposit his sperm sac. The male may even inject his 'intended' with a small amount of venom to subdue her.

After the male scorpion has laid down his sperm sac, he leads the female over it so that her abdomen makes contact with the sac and the sperm can enter her reproductive opening.

Mating can last from 1 to over 25 hours, with the male beating a hasty retreat when his task is complete — probably to avoid being eaten by the female (although sexual cannibalism is rare in scorpion species).

Scorpions are found all over Australia. Some inhabit forested areas, while others live in deserts. Desert species tend to be larger, more venomous and usually live in burrows deep in the ground.

The Little Marbled Scorpion is one of the most common Australian species. This scorpion is found right across southern Australia, in forests, hiding under rocks, fallen logs and behind bark, feeding on small insects. It has also been seen in urban houses, hiding in shoes and under clothing.

Scorpions feed at night on a variety of insects, spiders, centipedes, millipedes and even other scorpions. Some of the larger species eat small lizards and snakes. They take hold of their prey with their pincers and use their tail to repeatedly sting their victims. Death is almost instantaneous. Scorpions then dismember their victims with their pincers and eat the pieces, sucking out the juices and discarding the leftovers.

Scorpions can live from 4 to 25 years, depending on the species, and have been found in fossil records as far back as 425 to 450 million years ago.

Human victims say that being stung by a scorpion is like hitting your thumb with a hammer. The sting produces localised pain, numbness or swelling, sometimes lasting for a few days. Some people may be allergic to the venom of some species.

Applying a cold pack to the affected area helps relieve pain. Painkillers may be required in extreme cases. If the victim has an allergic response, such as breathing problems, or goes into shock, call an ambulance immediately.

Legs By the Hundred — Well, Almost

Centipedes and millipedes get their names because of their numerous legs. Both names come from the ancient Latin words *centium* (one hundred), *mille* (one thousand) and *pedis* (of the foot). So the names literally mean 'one hundred feet' and 'one thousand feet'. But these names are exaggerations. Most centipedes have between 40 and 80 legs. Millipedes can have as few as 60 legs or as many as 700.

Both centipedes and millipedes have poor eyesight (so they're not out to get you). But they both use their defence mechanisms against you if you disturb them: The centipede has tiny venom-filled claws and the millipede produces toxic secretions.

Centipedes

Centipedes are fast-moving and venomous little creatures, with many jointed legs. They're found in tropical and temperate zones throughout Australia. Tropical species can grow as large as 15–20 centimetres (6–8 inches) in length, but most are quite small, at 3–5 centimetres (1–2 inches).

Centipedes are *nocturnal* (active at night), feeding on insects such as flies, cockroaches and other small house pests. During the day they hide in crevices, in rotting wood or under rocks.

Some tropical centipedes are highly venomous and produce extremely painful bites. Thankfully the bite of the smaller centipedes living in temperate areas is usually much less painful.

The bite from a centipede is delivered from a pair of claws located just behind its head (see Figure 6-2). The centipede uses these claws to inject venom into its prey or to defend itself against predators. The forked tail that looks like it may deliver a sting is, in fact, harmless.

Very little is known about centipede venom. However, what scientists do know is that it can cause nausea, vomiting, dizziness, headaches and collapse. Usually, the most serious effect of a bite is severe pain. No deaths from centipede bites have been recorded in Australia.

Place a cold pack on the puncture wounds to reduce pain. In severe cases, generic painkillers, such as paracetamol, can help with pain management. If dizziness, headache, nausea or vomiting occur, seek medical attention immediately.

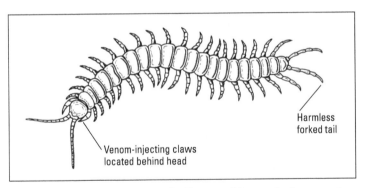

Harmless
forked tail

Venom-injecting claws
located behind head

Figure 6-2: Centipedes have one pair of legs on all but one body segment.

Millipedes

Millipedes (see Figure 6-3) don't bite or sting. Their main means of defence is to curl up to protect themselves. However, in addition to this, they secrete a toxic liquid that's designed to burn the *exoskeleton*, or outer skin, of ants and other insects. This secretion can burn larger predators' eyes.

Estuarine (Saltwater) Crocodiles have no fear of humans: Salties don't chew their prey, they crush it with their powerful jaws.

The colour of the deadly Eastern (Common) Brown Snake is variable.

The Eastern (Mainland) Tiger Snake is found in some of the most heavily populated regions of Australia.

The Coastal Taipan is the third most venomous snake in the world and feeds exclusively on mammals.

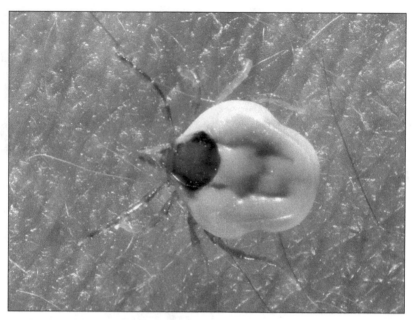

The Paralysis Tick is a tiny but deadly bloodsucking creature.

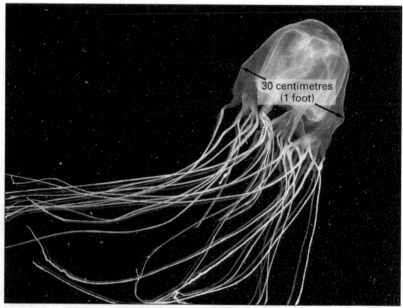

30 centimetres
(1 foot)

The Australian Box Jellyfish is the deadliest jellyfish in the world.

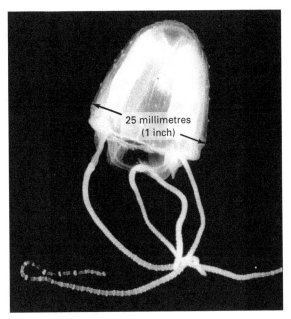

25 millimetres
(1 inch)

The Irukandji is a tiny transparent deadly jellyfish, small enough to get through stinger nets in the tropics.

The Great White Shark grows up to 7 metres (23 feet) in length, with triangular teeth up to 8 centimetres (3 inches) long. A single bite can be lethal.

The blue-ringed octopus is about the size of a golf ball. When provoked, it changes colour from dull brown to bright yellow with distinctive bright blue rings.

The beautiful patterns on cone shells make them tempting to pick up. But doing so could be fatal.

The Reef Stonefish is reputedly the most venomous fish in the world: The venom in its spines remains active for days, causing excruciating pain.

The Red-back Spider bites more people each year than any other Australian venomous creature.

The Sydney Funnel-web Spider rears up and strikes viciously with its venom-filled fangs.

The sting of a Honey Bee can be deadly. In Australia, more people die from the Honey Bee's sting than from any other species.

Bull ants have powerful jaws and a nasty sting in their tails.

The paper wasp delivers a potentially deadly sting when threatened.

Millipedes as insect repellent

Interestingly, lemurs on the Island of Madagascar have learnt to rub their fur with the secretions from millipedes to repel insect pests.

Also, for self-defence, the North American Wood Millipede produces a secretion to repel Fire Ants. This is currently of interest to scientists searching for a natural product that can be used on human skin against Fire Ant attacks. (For information on Fire Ants, see Chapter 15.)

The millipede's secretion is relatively harmless to most humans, producing only mild skin irritation and some discolouration. But people can experience pain, itching and swelling, even blisters and cracked skin, if they're allergic to the secretion. Also, if the secretion gets into the eyes, they can become irritated and sore, leading to *conjunctivitis* (an inflammation of the skin that lines the eyelid).

 Flush the affected area thoroughly with lots of water, and then use a cold pack to relieve itchiness and swelling. Get medical attention if swelling spreads or pain persists. If your eyes show signs of being affected, flush them well with water and visit a doctor immediately.

 If you accidentally touch or handle a millipede, you can avoid eye irritations from the secretions by washing your hands in warm soapy water.

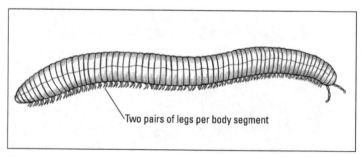

Two pairs of legs per body segment

Figure 6-3: The millipede. Most are less than 5 centimetres (2 inches) long.

Bloodsucking Bandits

Quite a lot of little critters survive by drinking blood. Some, like leeches and ticks, will drink any creature's blood. Others, such as some mosquitoes and lice species, only want the human variety.

In the following sections we cover the bloodsuckers that people really love to hate: Leeches, mosquitoes and ticks. Bloodsucking March flies, stable flies, sand flies (or midges) and black flies are mentioned in Chapter 10. We cover lice, bedbugs and fleas in Chapter 15.

Leeches

Leeches are bloodsucking worms. Their size and shape can vary greatly, depending on when and how big their last meal was. A leech can engorge to many times its normal size on one feed of blood. Australian leeches vary in size from 7 millimetres in length to 200 millimetres (0.25–8 inches) long when fully extended, and can range in colour from black or brown to maroon. They have a powerful sucker at each end of their body, as shown in Figure 6-4, and move by gripping a surface with their front sucker (at the head end) and bringing the rear sucker (at the 'tail' end) up to it, repeating this action to inch along.

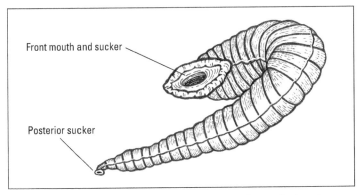

Figure 6-4: Bloodsucking leeches have powerful suckers at both ends.

Leeches latch on in different ways, depending on their subspecies. Here are the main groups:

- **Land or jawed leeches:** Jawed leeches have tiny teeth, which they use to latch onto a host. They also secrete *hirudin*, an anti-blood-clotting substance, so that they can feed until satisfied. (We tell you more about how hirudin works later in this section.)

  These are the leeches that attach to you while bushwalking.

- **Jawless or freshwater leeches:** These leeches insert a needle-like protrusion called a proboscis into a victim before secreting hirudin to dissolve blood clots as they form, so that the leech can continue feeding until satisfied.

  You're likely to encounter these leeches when swimming or wading in a dam, creek or river. They can also transfer from a freshwater fish to a fisherman!

- **Worm leeches:** No jaws or teeth here — these leeches swallow their prey whole. Their food consists of small insects and worms, not humans.

Come mealtime, most bloodsucking leeches have their 'favourites'. Some prefer mammals, while others get their blood from fish, frogs, reptiles, birds or even other leeches. But if their preferred host can't be found, they're quite happy to feed on human blood.

Getting attached

When a leech attaches itself to your skin, its bite usually goes undetected because you feel little or no pain. This is because it introduces hirudin to the wound. *Hirudin* is an anti-coagulant that stops your blood from clotting so that the leech can feed easily. When it has had its fill, the leech drops off.

Substances that prevent the clotting of blood are called *anti-coagulants*. These substances enable leeches, mosquitoes, ticks, fleas and other blood suckers to drink the whole of their blood meal in one bite. Many snake venoms also contain anti-coagulants. This ensures the venom spreads throughout the whole body, rendering the victim defenceless.

Most leeches have three jaws that inflict a Y-shaped puncture wound (although the Australian land leech has only two jaws and therefore makes a V-shaped wound).

Giving leeches the flick

Follow these steps to remove a leech from your body and then to treat the wound:

1. **Remove the leech by sprinkling salt, sea water or vinegar on the body of the leech, or use the heat from a recently extinguished match to force the leech to curl up and detach itself.**

   Do *not* pull a leech off, because parts of the jaw may remain and set up an infection.

2. **Wash the wound with water and apply pressure if the wound is still bleeding.**

3. **Use a cold pack to reduce pain or swelling.**

4. **Monitor all wounds for a few days.**

   Bacteria transferred from a leech bite can cause wound infection. If you notice an infection (continued redness and swelling, accumulation of pus and failure to heal), consult a doctor for appropriate treatment.

   Note that you may also experience delayed irritation and itching after a bite.

Allergic reactions to leech bites

Some individuals show an allergic response to leech bites. If the following occurs, seek medical advice immediately:

- Red blotches or body rash
- Swelling in parts of the body not near the bite (for example, around the eyes, face and neck)
- Dizziness
- Breathing difficulties

Leeches and doctors

Leeches were commonly used in blood letting, a popular medical practice up until the 19th century. *Blood letting* involved drawing the blood from the veins of a sick patient using leeches in the hope that the 'bad' blood was removed, thus curing the patient. The practice, which usually only weakened the patient, has been abandoned, except for a few specific conditions.

Today, leeches are used to help heal bruises, to treat black eyes,

and by plastic surgeons dealing with difficult skin and muscle flap attachments.

A leech can only be reused on the same patient.

Also, *hirudin*, the chemical secreted by leeches to prevent blood clotting, is used in the treatment of middle ear inflammations and is experimented with in preventing blood clots during blood sampling.

Mosquitoes

Mosquitoes transmit many diseases from person to person or from other animals to humans. Of the 300 different species of mosquitoes found in Australia, only a few are dangerous to man, and only the females of those species.

Females require a protein feed of blood before laying eggs, and your blood or that of another mammal will do.

Making a meal of blood

Female mosquitoes find their hosts by smell, carbon dioxide emissions, movement or heat. After locating a victim, the female pierces the skin, injecting a small amount of saliva at the same time (see Figure 6-5).

The female mosquito's saliva contains chemicals that prevent the victim's blood from clotting and can also contain viruses, bacteria or other microbes that can cause harm. After her fill of blood, the mosquito flies off to digest her meal. Later, she lays her eggs as a raft on water. Here, the eggs hatch and develop until they're ready to take flight themselves.

Viral diseases that can be transmitted to humans by mosquitoes in Australia include:

- ✔ **Australian encephalitis, also known as Murray Valley encephalitis and Kunjin virus:** This disease causes mild to severe damage to the nervous system. The damage can be temporary, permanent or fatal, depending on the severity.

    Cases of Australian encephalitis occur mainly in northern Australia (Queensland, Northern Territory and northern Western Australia).

- ✔ **Dengue fever:** This disease causes symptoms ranging from mild fevers to kidney failure and severe blood loss from ruptured blood vessels.

    In Australia, mosquitoes carrying the Dengue fever virus are restricted to northern Queensland.

- ✔ **Ross River virus:** Symptoms vary from fevers with or without a rash, to arthritis that can last from months to years.

    Ross River disease occurs in all states of Australia.

- ✔ **Barmah Forest virus:** Symptoms include fevers, rashes and joint pain.

    Barmah Forest disease occurs in most states of Australia.

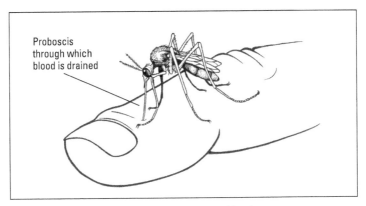

Proboscis
through which
blood is drained

Figure 6-5: A female mosquito dining on human blood.

Malaria

Malaria, caused by the microbe *Plasmodia*, is transmitted by the bite of the female Anopheles Mosquito. Victims suffer attacks of chills and fever, with 1–2 million people worldwide dying of the disease each year. The Anopheles mosquito has been officially eradicated from Australia since 1881. However, approximately 700–800 malaria cases are reported in Australia each year by travellers infected elsewhere.

The risk of transmitting malaria from one of these 'imported cases' to another person is virtually zero. For it to occur, a female Anopheles Mosquito would have to bite one of the few malaria victims available, then bite a healthy person. In Australia, once detected, malaria victims are hospitalised or at least bed-ridden and hence the risk is so minimal it's said to be 'virtually zero'.

Diagnosis of Dengue fever, and Australian encephalitis, Ross River and Barmah Forest viruses is made using a blood test.

Taking precautions

Mosquitoes can be a major nuisance in the workplace, during recreational activities and on social outings, like picnics.

When visiting an area known to have mosquitoes, preventing bites is much better than getting bitten. Simple precautions, like wearing long-sleeved shirts and long pants, using an insect repellent on exposed skin (provided you're not allergic to it), and scheduling outdoor activities to avoid dusk, when mosquitoes are most likely to be active, are helpful in avoiding bites.

At home, exterior doors with screens, windows with metal mesh or flywire and using bed nets help keep mosquitoes out. Insect coils and electric insect zappers are useful in keeping mosquitoes at bay around the outside of the home.

Covering water tanks with mesh or flywire and emptying all garden containers, tyres, roof guttering and tins of water prevents mosquitoes from breeding around your home.

When camping, use a tent that has netting on entrances and window flaps to prevent night-time bites.

Mosquito bites

In most individuals, a mosquito bite causes swelling, redness and itchiness. Some people suffer more severe swellings.

A small number of individuals may even have a severe allergic reaction (anaphylaxis) to a mosquito bite. Symptoms of the allergic reaction can include large swellings or hives and itchiness at the bite site, stomach cramps, difficulty breathing and swallowing, coughing, wheezing and fainting. First aid for anaphylaxis is described in Appendix A.

Wash bites with an antiseptic solution. Cold packs can help relieve pain and swelling. Calamine lotion or a specific product for treating insect bites (available at pharmacies and major supermarkets) can help ease the itchiness.

Avoid scratching bites — this can lead to infection.

Call an ambulance if the bites develop into large hives or big red swellings, or if the victim experiences breathing problems, dizziness or severe pain.

Tiny ticks

Ticks are bloodsucking, external parasites that need a blood meal at each of their four stages of development. Ticks have been responsible for at least three human deaths in Australia since 1979.

A *parasite* is an animal or plant that gains its nutrition from another plant or animal rather than finding its own food. Some parasites, such as tapeworms, are internal and live inside their hosts. Others latch onto the outer skin of an animal (or person). This group includes ticks and leeches (we cover leeches earlier in this chapter).

Of the approximately 75 species of ticks in Australia, the Paralysis Tick (shown in the colour section) is the most dangerous by far. It inhabits a 20 to 30 kilometre-wide (12 to 19 mile) stretch of land inland from the coast along the entire eastern seaboard of Australia. This area is also where the majority of Australians live, so encounters with these parasites are relatively common.

The Australian Paralysis Tick is found in humid, moist, bushy areas, all year round in Queensland and northern New South Wales, and from spring to autumn in southern New South Wales, Victoria and Tasmania.

Ticks rely on passing animals for a feed and to transport them. They go 'questing' for a host by climbing bushes and grasses, then waving their forelegs slowly in the air, hopefully making contact with a furry animal or human.

Both female and male ticks quest for a host — the female for a blood meal; the male to find a female in order to mate. Male ticks rarely blood-feed on a host. The adult female Paralysis Tick, shown in the colour section, will feed for up to 10 days, drop off the host and then lay eggs on nearby foliage over several weeks.

The consequences of tick bites

Most cases of tick bite have little or no serious effect. The toxin injected during the bite usually causes only a local irritation or a mild allergic reaction. However, some bites can result in life-threatening illnesses, including paralysis (especially in children), tick typhus and severe allergic reactions. The wound can develop a black dry scab.

Symptoms of tick paralysis include fever, grogginess, headache, rashes, swollen and tender glands, unsteadiness when walking, dislike of bright light and partial facial paralysis. Tick paralysis begins slowly and intensifies as the tick engorges during feeding. This takes several days. Even after the tick is removed, the patient's condition can continue to deteriorate for a time. Recovery can be slow.

Symptoms of tick typhus include headache, rashes, swollen glands, fever and flu-like symptoms.

Tick typhus, also known as Spotted Fever, is an infection caused by a bacteria-like organism that is transmitted from the blood of other animals to humans by ticks, fleas, lice and mites.

Symptoms of allergic reactions to ticks

Reactions vary from mild irritation with localised swelling, to pain and widespread swelling, to severe and life-threatening anaphylactic shock — the victim feels faint, grows pale and develops breathing difficulties.

These reactions can occur with any tick stage bite (be it larval, nymph or adult). People who develop severe allergic reactions must *always* avoid contact with ticks and steer clear of areas potentially infested with them.

Preventing ticks from biting

No one wants to pick up a Paralysis Tick. Here's how best to avoid doing so:

- ✔ Wear long-sleeved shirts, long pants tucked into socks or gaiters and a wide-brimmed hat to reduce the likelihood of being bitten by a tick.
- ✔ Spray yourself and your clothing with insect repellent to discourage them attaching when you're outdoors.
- ✔ Dress in light-coloured clothing. This helps you to spot and remove a tick easily.
- ✔ Remove your clothing after you return from a tick-infested area or after handling any animal that could host ticks. Put your gear into a hot clothes drier for 20 minutes to kill the ticks and prevent a tick infestation in your home.
- ✔ When searching your body for ticks, look behind your ears, on the back of the head, in neck and hairline creases and in the groin area, armpits and behind your knees.
- ✔ Check your children and pets, too. (Dogs and cats can die from tick paralysis.)

Ticks can infest domestic, feral and native animals. So always wear gloves when handling a sick or injured animal, and inspect yourself for ticks afterwards.

Tick treatment

Don't muck around if you suspect someone (or yourself) is having a severe allergic reaction or if you find a tick on your body.

- ✔ **Treating anaphylactic shock.** The victim might faint, turn pale and develop breathing difficulties and you must call an ambulance immediately. Be prepared to give mouth-to-mouth resuscitation and external heart massage until the ambulance arrives.

- ✔ **Removing a tick found on your (or someone else's) body.** You need to get the tick off as soon as you find it. Use tweezers to grab the tick, making sure to grasp it firmly with the tweezers flush against your skin. Gently but firmly pull the tick out. (You don't want to leave any part of the tick behind.)

- ✔ **Dealing with an infestation of larval ticks.** Larval ticks look just like adults, but they're smaller and paler. They infest in groups because after an adult female lays a batch of 2,000 to 6,000 eggs (20 to 200 per day — yuk!) only a small proportion hatch, but all at about the same time. If you're stricken with this, you'll notice several tiny pale ticks clustered together. Treat an infestation by soaking in a bath with 1 cup of bicarbonate of soda dissolved in it, for 30 minutes. (You can buy bicarbonate of soda at the supermarket.)

After you've removed a tick, keep it in a glass jar for four days. If, during this time, you develop any of the symptoms of paralysis, typhus or allergic reactions (refer to 'The consequences of tick bites' earlier in this section), take the tick with you to the doctor for ID purposes.

An antivenom, developed for treating tick paralysis in animals, is also used to treat humans affected by tick paralysis.

Chapter 7

Four-Legged Friend or Foe?

*O*ddly enough, most of the earth's dangerous land creatures are smaller than humans — and some, like bees, wasps, ticks and spiders, are much, much smaller. Most deadly land creatures have more legs than we do — or no legs at all.

However, a few four-legged mammals can present a serious danger. The Dingo, along with other wild dogs and feral animals, such as pigs, goats, horses, donkeys, cattle and cats, are attracted to human settlements and camp sites where food scraps, livestock and pasture abound. Bats, also mammals with four legs (two of which they use more like arms), can be found just about anywhere and can transmit some serious diseases to humans. Looking beyond mammals and into the world of reptiles, some of Australia's native goannas are now classified as venomous.

Many of the four-legged animals we present in this chapter don't retreat when threatened by human presence (as most snakes and spiders do). Like people, they'll fight to defend their territory. Even the very shy Tasmanian Devil will bite ferociously to defend itself and its food.

Dingoes: Wild Dogs Down-Under

Dingoes are an Australian breed of wild dog, thought to have been brought to Australia from Asia 6,000 to 10,000 years ago by Asian sailors. The name *Dingo* was first used by the early European settlers of New South Wales to describe the dogs belonging to the local Aboriginals.

Dingoes (see Figure 7-1) are commonly yellow-brown or ginger in colour, with white feet and a white tip at the end of the tail. Dingoes with other colourations, such as black and tan, or brindled, are the result of cross-breeding with other wild or domestic dogs. When compared with dogs, Dingoes are of medium size, standing 50–60 centimetres (20–24 inches).

Dingo country

Wild Dingoes are seldom found in heavily populated areas, such as in cities and towns; instead, they live in farming and outback areas across most of Australia. They can breed with wild or feral dogs, and so many Dingoes on the mainland of Australia are mixed breeds. Because Fraser Island is isolated from the mainland (located near southeast Queensland), the population of Dingoes found there is of pure breed. Dingoes are also a popular exhibit in Australian zoos.

Figure 7-1: The Dingo: The wild dog of Australia.

The longest fence in the world

A 5,400-kilometre-long and 2-metre-high (3,350-mile-long and 6.5-foot-high) wire mesh fence is designed to protect the sheep grazing in the southeast region of Australia from Dingoes. This Dingo Fence, also known as the Wild Dog Fence, stretches from the Great Australian Bight in South Australia to about 150 kilometres (93 miles) short of the Sunshine Coast in Queensland. It was built more than 120 years ago as a rabbit-proof fence, but was modified in 1914 to keep Dingoes out. Although the Dingo Fence hasn't been completely successful in stopping Dingoes from reaching pastoral regions and sheep, it has kept other threats, including brumbies (feral horses) and feral pigs, in check.

Dingoes on the attack

In the wild, Dingoes eat whatever food is available, from kangaroos to insects. Living in groups or packs of 2 to 10 in a well-defined territory or range, they can hunt individually or cooperatively when after large prey, like kangaroos or wallabies. Dingoes are a threat to livestock, particularly sheep. But packs of Dingoes will also kill young cattle. Stories of Dingoes attacking humans have surfaced since early settlement in the 1780s.

Recent attacks have occurred in areas where people feed Dingoes, which causes the animals to lose their fear of humans and aggressively seek out food. Dingoes that have become accustomed to being fed by humans will steal food from camp sites. When camping, you need to think carefully about food storage. Dingoes can overturn ice chests, push over bins and tear open bags to get at food. They sometimes rip tents open to seek food and may even consider humans as food — especially small children. The tragic Azaria Chamberlain case illustrates this. At 9 weeks of age, she was taken from a tent in central Australia by a Dingo, and never seen again.

Many Dingo attacks have occurred on Fraser Island, including a tragic fatal attack on a 9-year-old boy by two Dingoes in 2001.

To be safe from Dingo attack:

- ✔ Never feed Dingoes or leave food out for them.
- ✔ Store food securely. Burn scraps or use the lidded bins provided at camping grounds.
- ✔ Never walk alone and always make sure children are accompanied by an adult.

If threatened by a Dingo:

- ✔ Don't run or wave your arms about. Stand tall, facing the Dingo, keeping your eyes on the animal. Fold your arms across your chest, and if you're with two or more other people, stand back to back.
- ✔ Gradually back away.
- ✔ Don't scream. Call for help using a loud, strong voice.

If attacked by a Dingo, fight back. Kick or lash out at the Dingo with a stick, spade or whatever is at hand.

Dingo bites can inflict deep, jagged, bacteria-laden wounds. Be sure to thoroughly clean any wound using an antibacterial solution and warm water, allowing the water to run over the region for several minutes. Then dry and bandage the wound.

See a doctor to find out if you need a tetanus shot, or if the wound is large and requires stitching, or if it becomes infected. For more information on dealing with a medical emergency, see Appendix A.

The Mighty Water Buffalo

Two breeds of water buffalo live in Australia: The River Buffalo from western Asia, with curled horns, and the Swamp Buffalo from eastern Asia, with swept-back horns (see Figure 7-2). The Swamp Buffalo is the more common of the two. Both breeds are found in the tropical wetlands of the Top End of Australia — the northern part of the Northern Territory, the Kimberley Region of Western Australia and the Gulf country of Queensland.

River Buffalo

Swamp Buffalo

Figure 7-2: Water buffaloes. The shape of the horns identifies the breed.

At home in the wetlands

Water buffalo were first imported to Australia in 1826 from Timor, to supply meat for the remote settlements on Melville Island and in the Northern Territory and Queensland. When these settlements were abandoned in the 1940s, the buffalo were set free. Their ability to live in native swamplands and flood plains allowed them to become a major pest across the Top End in the 1960s. They were destroying wetlands and were known carriers of the diseases brucellosis and tuberculosis, which affect native species and livestock. A successful eradication campaign, developed to protect the meat export industry, has culled feral buffalo numbers back to manageable levels.

Back to the farm

For many years now, the farming of redomesticated herds of buffalo has supported several industries. For example, because buffalo meat is lean and low in cholesterol, it's suitable for human consumption, and is sold locally and overseas.

Buffalo meat is also used for pet food, whereas buffalo milk is used to manufacture specialty cheeses and yoghurts.

These animals are also hunted as game, and their horns are mounted and used for trophies.

Attacks on humans

Domesticated water buffalo are placid by nature and can be treated like most cattle. However, wild water buffaloes can attack humans if caught unawares and frightened. They also run and charge to defend their calves. Like a bull in a ring, they will chase a person they see as a threat, using their horns to attack and gore. A charging water buffalo is quite capable of killing a human.

When in wild buffalo country:

- ✔ Never walk towards a group of buffalo. Make your presence known by walking in the open with other people, making a medium amount of noise.
- ✔ At night, carry a torch and make some noise; you don't want to startle the buffalo.

Being gored or trampled by wild water buffalo can cause major injuries. Stem any bleeding by applying pressure to the wound with any material at hand (a towel, shirt or coat). If the injury is on an arm or a leg, raise the limb to slow blood loss. Call an ambulance.

Farm Animals Gone Wild: Ferals on the Loose

When Europeans first settled in Australia, they brought with them domestic farm animals from England and other parts of Europe to stock farms and provide much needed food and fibre. When these domesticated breeds escaped into the wild, they became feral — wilder, fiercer and more unpredictable versions of their breed. Domesticated animals, including pets, continue to get out and go feral in the bush, and have become a big problem in Australia. Because most of these animals have no natural predators in Australia, they're able to out-compete native animals for food, shelter and space. They also cause soil degradation, erosion, water contamination and vegetation destruction.

Not all feral animals pose an immediate danger to humans, but indirectly they affect livelihoods by competing with farm and native animals for resources, such as food and water, by destroying fences and crops, and by spreading diseases that may end up in domestic herds.

Pigging out: Feral pigs

Early settlers on the Australian mainland often allowed their pigs to roam free, so before long these animals became feral pests.

Because pigs eat almost anything, they can live just about anywhere. Feral pigs roam western Victoria, most of New South Wales and Queensland, and the Top End, ranging from Cape York in Queensland to the Kimberley Ranges in northern Western Australia — anywhere they can find a sufficient supply of water. Feral pigs are absent from inland arid areas because of the lack of water, and they're not found in Tasmania.

Most feral pigs are black, but other colours have been introduced with cross-breeding. They have coarse hair, tusks and a solid build, though they're somewhat smaller and narrower than many domestic breeds (see Figure 7-3). They look meaner, too.

Stay well clear of any feral pig you meet. Where there's one, there are usually more; herds can be a hundred strong. They're very intelligent, have a powerful sense of smell and are aggressive towards humans. They sometimes chase and attack people. Feral pigs also carry the diseases Murray Valley encephalitis and Ross River virus.

Pigs in the bush

Feral pigs destroy natural habitats. These pigs have a habit of wallowing in mud and trampling areas around waterways, then destroying vegetation as they search for food. This deprives native animals of food and nesting sites and causes erosion of river banks.

Feral pigs also damage fences and cereal crops, eat young lambs, carry diseases that can be transmitted to humans and have the potential to carry foot-and-mouth disease and swine fever (were these diseases to enter Australia).

Figure 7-3: Feral pigs have been known to chase and attack people.

To discourage feral pigs from invading camp sites:

- ✔ Never feed feral pigs or leave food out for them.
- ✔ Store food and scraps securely — pigs will eat almost anything. Burn scraps or use the lidded bins provided to dispose of rubbish.
- ✔ Never keep bait or fish scraps near your camp site. These attract feral pigs.
- ✔ Inform rangers if you see any feral pigs.

If you're chased by a feral pig:

- ✔ Climb the nearest tree and wait it out.
- ✔ If climbing something isn't an option, sit on the ground and use your feet to kick the pig away — or hit out with a branch.

Feral goats: No kidding

Compared to other livestock, such as cows and sheep, goats are pretty versatile — providing meat, milk and fibre in a small and compact sturdy body that eats almost anything. Goats were officially brought to Australia by the settlers on the First Fleet in 1788. And goats being goats, it wasn't long before they escaped into the wild and established feral populations on the mainland.

Up to the mid 1800s goats were deliberately released on islands across the Pacific for an emergency supply of food for marooned sailors. Even Captain James Cook did this during his 1768–1771 voyage to the Southern Hemisphere. He dutifully released goats and rabbits in New Zealand and, when he found Australia, on the mainland.

Feral goats — around two million of them! — live in all states of Australia (except Tasmania). They prefer semi-arid regions with rocky terrain. The Dingo is one of the goat's few natural predators, so these ferals are seldom found in Dingo country. They avoid rainforests and wetlands (too wet), and deserts (too dry).

Feral goats destroy native vegetation, allowing erosion and preventing regeneration. They also compete with native animals for food, water and shelter, and with domestic stock (sheep and cattle) for pasture, particularly during drought.

Feral goats are subject to footrot, and because of their liking for pasture, they're a reservoir of this disease, spreading and reinfecting domestic livestock. And feral goats, like feral pigs, are potential carriers of foot-and-mouth disease, if an outbreak were ever to occur in Australia.

Feral goats can breed twice a year, often producing twins and triplets, making them a difficult feral pest to eradicate. Feral goats can weigh up to 40 kilograms (88 pounds) and grow to 90 centimetres (35 inches) in height, with shaggy black, brown, white or mottled coats. Both males and females have permanent horns.

If you're chased by a feral goat, keep an eye on the goat at all times and watch out for those horns:

- ✔ Seek protection behind a tree trunk or fence and call for help. (A car horn or other loud noise will often distract a goat and cause it to flee.)

- ✔ Grab a branch or something else to use as a weapon, particularly if you can't make it to the nearest tree or find a fence.

Feral cattle, horses and donkeys

Feral cattle, horses and donkeys inhabit many remote outback areas. They sometimes trash and run through camp sites, injuring people and damaging personal belongings and camping gear.

Feral cattle

Feral cattle can be traced back to their ancestors, who were introduced to Australia in the interests of farming. In many remote areas of Australia (the Outback), where paddocks have little or poor feed, cattle grazing goes unchecked for many months at a time. So, it's not surprising these animals escape and become feral.

Feral horses and donkeys

Horses arrived with the First Fleet (along with early settlers, convicts, government officials and military personal) in 1788. Since much of the land was first grazed with little or no fencing, it didn't take long for horses to escape into the wild. And by the 1830s, bush horses or *brumbies*, as Australians call them, were well established in the scrub surrounding Sydney.

Brumby is an Australian word used to describe a wild, unbroken horse. The brumby is romanticised in Australia's most well-known bush poem *The Man from Snowy River*, by A B (Banjo) Paterson, which tells the tale of a colt that escapes from grazing land to run with the wild horses in the Australian Alps. The poem was made into a successful movie in 1982. Although the brumby is regarded as an Australian icon, the thousands of brumbies that still run wild today are considered to be a pest by many, causing erosion and polluting lakes, rivers and creeks.

Donkeys were introduced into Australia in 1866 to serve as pack animals, and by the 1920s feral herds were reported.

Today, Australia is home to more than 300,000 feral horses and up to 1.5 million feral donkeys, mainly in central and northern Australia. Both brumbies and feral donkeys destroy native vegetation, causing soil erosion — all hooved animals do; Australian native animals are all soft-footed. They compete for food with native animals and livestock, destroy

Bali Banteng

The Banteng, an Indonesian breed of cattle, was introduced to Australia in 1849 for farming. Twenty cattle were imported to a farm on the Cobourg Peninsula, part of Arnhem Land in the Northern Territory. When farming in the area failed, the cattle were left to fend for themselves. Conditions proved to be satisfactory for the Banteng and their population grew and started to spread to the south. Fencing across the base of the peninsula has contained the cattle, and authorised hunting keeps the numbers in check while making some income for local Aborigines.

In its native country of Indonesia, the Banteng has been crossbred with other breeds. Australia is now considered the only country that has purebreds.

burrows and ruin water supplies for native animals, and they spread weeds via their dung, manes and tails.

Never attempt to feed or pat wild cattle, horses or donkeys. No matter how tame they may appear to be, they carry ticks (refer to Chapter 6) and diseases that can be conveyed to you if they bite.

Camping in well-established, fenced-off camping sites is one way to prevent being trampled by spooked herds.

Ferocious ferals (feral cats and dogs)

Feral cats and dogs (see Figure 7-4) drastically affect the survival of many of Australia's smaller native animal species by competing with them for food and shelter. They also harbour diseases that affect native species, livestock and humans. Feral cats and dogs can be very aggressive, especially if cornered. Because they look like domestic pets, they can lead people into a false sense of security about handling them. But don't be fooled, they're dangerous!

Feral cats

Cats were brought to Australia with European settlement from 1788 onwards and may have arrived even earlier, from Dutch shipwrecks in the 17th century.

Feral cats are found all over the Australian mainland, Tasmania and on many offshore islands. Subject to cat flu (or influenza) they can't survive in the extreme wet, so they tend not to be found in rainforests and snow fields.

Feral cats feed mainly on small mammals, and can survive on very little water. They also eat birds, reptiles, amphibians, fish and insects, if need be. They can breed in their first year and can produce two litters a year. Natural predators include Dingoes, foxes and Wedge-tailed Eagles. Their diverse feeding habit, lack of predators and high breeding rate make these ferals particularly hard to eradicate.

The impact of feral cats on native fauna is significant: Scientists hold them responsible for the extinction of the Red-fronted Parakeet on Macquarie Island (in Australia's Antarctic region), and in all probability for the extinction of many other small mammals on the mainland. They have seriously affected bilby and numbat populations, and pose a threat to ground-dwelling birds, such as Australia's lyrebird. If rabies were accidentally introduced into Australia, feral cats would accelerate its spread to the human population.

Feral dogs

The first feral dogs were, of course, the Dingoes (discussed in detail at the beginning of this chapter), but early European settlement brought other breeds, and when these strayed from home and became feral, they added to the wild dog problem.

Figure 7-4: Feral cats and dogs.

Feral dogs are a major problem in sheep country and when cattle are breeding. Hunting in packs, they can single out a sheep or a cow, especially if it's weakened by sickness or age, and kill and eat it. They also savagely defend their kill and themselves, and will attack others (including humans) if necessary.

Avoiding nasty feral cat and dog encounters

Here are some guidelines to help you steer clear of being on the receiving end of a nasty bite or scratch from a feral cat or dog:

- ✔ Never feed feral cats or dogs and don't leave food out for them. Feral cats and dogs like to feed from camp sites and rubbish bins, so ensure you place your left-over food scraps and rubbish in a lidded bin or burn them.

- ✔ Don't try to catch a feral animal yourself; always inform a ranger (if camping) or the local council if you see any feral cats or dogs. These people know how to deal with them.

- ✔ If you accidentally corner a feral cat or dog, back off slowly and give the feral animal a means of escape; leave catching the animal to the experts!

First aid, after an attack

Wounds caused by feral animals can be particularly varied and nasty:

- ✔ Larger feral animals can inflict serious injuries, including severe bites, bruising and broken bones that may require urgent medical attention. (See Appendix A for more information on dealing with an emergency.)

- ✔ All feral animals can transfer bacterial infections when they bite or scratch — particularly the feral cat — which can lead to illness within a few days.

To treat minor wounds, clean the affected area thoroughly with soap (or an antibacterial solution) and warm water to remove dirt, saliva and bacteria. Allow the water to run over the wound for several minutes to make sure it's clean, then dry and bandage. Seek medical help if the wound needs stitching or if your tetanus immunisation is out of date.

Keep a close eye on dressed wounds over the next few days for signs of infection (redness, swelling, weeping from the wound, pus, increased pain). See a doctor if symptoms develop.

Face to Face With the Devil: The Tasmanian Devil

The Tasmanian Devil (often simply called the Tassie Devil) gets its name from the way it looks and sounds when it's energetically scavenging for food: The Tasmanian Devil grunts and snuffles and growls, and lets out the most unearthly screams. At the same time, its ears turn red as extra blood flows to them, and the animal displays its big, sharp pointy teeth in its wide pink mouth. The Tasmanian Devil's jaw has the biting power of an animal four times its size. Despite appearances, though, the Tasmanian Devil is usually fairly timid.

Tasmanian Devils are thickset, dog-like animals weighing about 6 to 8 kilograms (13 to 18 pounds). They have a powerful head and thick black coat with a white collar or patch around their neck (see Figure 7-5). They stand about 30 centimetres (1 foot) in height and carry their young (up to four joeys) in a backwards-facing pouch. They inhabit coastal scrub and forests in Tasmania. Some historians say that Dingoes competing for the same food resources caused the Tasmanian Devil's extinction from the mainland.

Figure 7-5: The Tasmanian Devil has the biting power of an animal four times its size.

A devil of a swim

Tassie Devils like to get their feet wet and splash about. And they're quite good swimmers. But if they have young in their backwards-facing pouch, they only swim short distances. Tassie Devils often take to the water to cool off in summer, paddling about, rolling around or sitting down in the water. You can sometimes even see them washing their 'hands'.

The ultimate vacuum cleaner

Tassie Devils are the ultimate vacuum cleaners of the animal world; they can devour any part of a dead animal they can get their teeth into: Meat, bone, offal, feathers, fur — the works. Even though they prefer other creature's kill, they do hunt their own prey. In the wild, this may be snakes and lizards, bush rats and small wallabies. Close to settlements, they kill lambs and sheep.

Food: The Devil's first priority

Tasmanian Devils often fight each other over food. The older the Tasmanian Devil, the more battle scarred it will appear. Fighting also spreads a face tumour cancer among the species, which is presently endangering the survival of the Tasmanian Devil. During fights, tissue is exchanged between devils (like a transplant) enabling the cancer to spread to other Tasmanian Devils. This cancer hasn't been found in any other species, and is currently the focus of many research projects.

Tasmanian Devils aren't particularly dangerous to people. They won't attack you unless forced to defend themselves, their young or their food; especially their food. Their powerful jaws can inflict serious injuries on humans. They also carry ticks that can be transferred to people who come in contact with them.

If you're bitten by a Tasmanian Devil, clean the wound thoroughly with an antibacterial solution and warm water to remove dirt, saliva and bacteria. Repeat this twice to ensure all dirt and other matter is removed.

The fascinating reproduction of Tasmanian Devils

When born, Tasmanian Devil babies (joeys) are blind, deaf, hairless and the size of a grain of rice. Up to 50 are born at one time, but only a maximum of four survive, because each newborn has to crawl its way from the vaginal opening to one of four teats located in the backwards-facing pouch of the mother. They remain suckling in the pouch for up to three months. Then the joeys are left in the nest, inside the den, until the mother returns with food, for a further three months. At about six months of age, the joeys begin to venture out on their own in search of food. Many joeys don't make it to adulthood, often killed by other Tasmanian Devils in fights over food. If they survive adolescence, they can live for a further 6 to 7 years.

If the wound is minor (a simple bite with no tissue missing), dry and bandage the wound. Then check your body for tick infestation (refer to Chapter 6). If the wound is more serious, seek medical help immediately. Also consult a doctor about a tetanus shot or if signs of infection develop.

Mammals with Wings

Bats are the only mammals that can truly fly. Some possums can glide, but bats have front leg-like limbs supporting the flight membranes in their wings. Bats also use their wings to wrap around themselves when they're cold and to 'fan' themselves when they're hot.

Australia has more than 90 species of bat. They live in large colonies (groups) and are most common in tropical regions, but some long-eared species can be found everywhere from Cape York, Queensland, to Southern Tasmania. Australian bats fall into two main groups:

- Large bats that eat fruits. Fruit bats are often called flying foxes because of their fox-like facial features.
- Smaller bats that eat insects. They have large echo-locating ears and small snubbed noses.

The hidden benefits of bats

The Chinese believe bats are a sign of good luck! Other people interpret bats differently. For example, uses for bats include:

✔ A traditional food for Australian Aborigines.

✔ A herbal medicine to stave off colds.

✔ A cure for baldness. In some parts of Asia, bat dung is thought to cure baldness. (Imagine getting around with that stuff on your head!)

✔ Solomon Islanders use bat wings to make kites.

Like other mammals, bats have fine, dense fur and suckle their young on milk. Most species give birth to a single offspring, but some species can produce twins. The mother carries the newborn on her chest during rest and flight, until they become too heavy. Then the young are left in night nurseries while the mother looks for food. When the young can fly, they join their mother's nightly searches for food.

Nature's little helpers

Fruit bats use their keen sense of sight and smell to locate their diet of fruit and flowers. These bats play an important role in seed dispersal and pollination of native plants in the environments in which they live.

Figure 7-6: The two types of Australian bat — a fruit bat (left) and an insectivorous bat (right).

The smaller insect-eating bats rely on echo-location rather than their sight to find food. They also play an important environmental role in keeping insect numbers under control. However, large groups of fruit bats can also ruin fruit crops, cause power blackouts and become a public health concern when they set up their camps in areas like Botanic Gardens.

Bat diseases

In 1996 and 1998, two people died after coming into contact with bats carrying a virus called Australian Bat Lyssavirus (ABL). When bats were tested for this virus, it was present in several species of fruit bats (and scientists warned that the virus could also make its way into the insectivorous bat population). The ABL virus is closely related to the rabies virus. Fortunately, the rabies vaccine and rabies immunoglobulin can be used to protect people against ABL.

An *immunoglobulin* injection is a dose of antibodies that fights disease-causing micro-organisms or particles. It doesn't provide future immunity to the disease, but works quickly to eradicate the present infection.

Transmission of the infection to humans is extremely rare. As a blood-borne virus, ABL can only be transmitted through a bite or scratch from an infected bat or by splashes of infected blood or urine into a person's eyes or nose. For this reason, bats (particularly sick or injured ones) shouldn't be handled unless the person is fully inoculated against rabies. If you do find a sick bat, call the local wildlife care group or contact your region's Parks and Wildlife Service (refer to Chapter 2).

If you're bitten or scratched by a bat, wash the wound thoroughly with soap and water for about five minutes and contact a doctor immediately so that you can be administered with a rabies vaccination.

If you get bat saliva in your eyes, nose or mouth, flush the area thoroughly with water and then seek medical advice.

Another viral disease, equine morbillivirus, is also carried by a small number of fruit bat species. Three people and several horses have died from this disease. At the time of publication, there's no evidence that equine morbillivirus can be transmitted directly from bats to humans.

Lizards: It's All a Bluff — Maybe

Some Australian lizards, like the Frilled Lizard shown in Figure 7-7, look very ferocious and hiss loudly if threatened. They can give you quite a scare if you encounter one unexpectedly. But they're not venomous and even a bite is unlikely to be dangerous to humans as long as the wound is carefully cleaned and treated with antiseptic lotion.

Until recently, most scientists believed that not one Australian lizard was venomous. In fact, only two lizards throughout the world were known to be venomous: The Mexican Bearded Lizard and the Gila Monster, both of which are native to the drier and warmer parts of North America. But in late 2005, a team of scientists at the University of Melbourne discovered that Australian goannas and the Bearded Dragon do have venom glands. Their venom is toxic enough to kill their prey, but is not believed to be deadly enough to kill healthy adult humans. However, this venom can be fatal if you're allergic to it.

Venomous goannas don't have fangs. Their venom is stored in their saliva. They must bite down heavily to inject their prey with sufficient venom to kill it.

About 20 species of goanna live throughout the Australian mainland — none is native to Tasmania. Australia's largest goanna is the Perentie Monitor (see Figure 7-8), which can grow up to 2.5 metres (8 feet) long.

Figure 7-7: The Frilled Lizard looks and sounds ferocious — but like most lizards, it's quite harmless to humans.

The Perentie Monitor lives in the outback of central Australia — places like Alice Springs and central Western Australia.

The Bearded Dragon (shown in Figure 7-8) is a goanna found in the Outback regions of the Northern Territory, Queensland, South Australia and New South Wales. Bearded Dragon venom contains toxins only previously found in rattlesnake venom. The Bearded Dragon grows to 25 centimetres (10 inches) in length. Other goanna species can be as little as 12 centimetres (5 inches) long.

If you get too close to a goanna, it will attack you by latching on with its teeth, biting down hard, and gripping you with its talons (claws). When biting, the goanna may keep squeezing with its jaw and twist its body to keep the pressure up. To release the goanna's jaw, firmly press into each side of its head where its mouth ends. The goanna will then let go.

Because goannas eat fresh prey and *carrion* (already dead animals), a goanna bite carries a very real risk of transferring bacterial infections. Bacteria can be driven deep into a wound by the goanna's talons or teeth. After an attack, cleanse the wound with an antibacterial solution and warm water, and consult a doctor for a tetanus shot or if the wound shows signs of infection (redness, swelling, weeping from the wound, pus, increased pain).

Figure 7-8: A Perentie Monitor, Australia's largest goanna (left), and the Bearded Dragon (right). Both goannas are poisonous.

Chapter 8

Watch the Birdie

Kookaburras and Emus are two of Australia's best known birds. Along with the kangaroo and Koala, these birds are living icons of this island continent. At first glance, neither the kookaburra nor the Emu is dangerous, but both have sharp beaks that have injured people. In this chapter, we also cover a couple of other unique Australian birds that can potentially injure humans. These include the

✔ Wedge-tailed Eagle, Australia's largest bird of prey and an awesome sight to behold

✔ Southern Cassowary, less known to Australians, due to its limited distribution in the rainforests of northern Queensland

Of course, many other native birds may attack people if provoked. For example, swooping birds that you're likely to find on or near the beach (such as plovers, gulls and terns) are described in Chapter 10. Other birds that you're more likely to encounter in suburban parks and gardens (such as Magpies) are covered in Chapter 14.

Australia's Cackling Birds: Kookaburras

Australia has two species of kookaburra: The Laughing Kookaburra (shown in Figure 8-1) and its tropical cousin, the Blue-winged Kookaburra. Kookaburras belong to the same family as forest kingfishers and have the same overall body shape, but kookaburras are larger and heavier in build. The most distinguishing feature of kookaburras, however, is their call. The raucous cackle of a kookaburra is easy to identify, even by the youngest Australians (we provide more information on distinguishing a kookaburra's call later in this section).

Kookaburras spend much of their time perched on low branches, patiently surveying the ground for prey — small snakes, lizards, mice and frogs. When they spot dinner, they gracefully swoop down with their beaks open to snatch the food from the ground or water. Small prey is eaten whole, but larger prey is killed by bashing it against the ground or a tree branch.

Kookaburras can live for up to 20 years and tend to reside in one area all their lives. Because of this, they become well known to local residents, and some people try to hand-feed them and train them to come when called. But no one can ever really tame a kookaburra, and attempts to do so can lead to accidental injuries.

Snacking on snakes

When kookaburras eat snakes they regurgitate the reptile's venom glands, bones and other potentially poisonous tissues as pellets of undigested food.

These pellets fall to the ground beneath their perch.

Kookaburras are quite shy and easy to overlook as they sit quietly on a branch overseeing their territory. Finding a scattering of these pellets beneath a tree branch enables birdwatchers to locate a current kookaburra perch and observe the kookaburra's habits.

The Laughing Kookaburra

The voice of the Laughing Kookaburra is probably the most recognisable bird call in Australia — a long and high-pitched cackle, a loud 'koo-koo-koo-koo-koo-kaa-kaa-kaa', often started by one kookaburra, with others joining in. This bird's natural habitat is the open eucalypt forests of Australia's eastern and southern states, including Tasmania. The Laughing Kookaburra has also been introduced to forests in the southwest region of Western Australia.

 An Australian Aboriginal legend says that the kookaburra laughs to herald in a new day. One folktale suggests the kookaburra laughs, as do people, when in a happy mood. Another folktale says the kookaburra is laughing at the folly of people having to work so hard to find food. Other theories suggest that the kookaburra laughs to warn of forthcoming danger or rain. Whatever the reason for its cheerful chuckle, the kookaburra's call is like no other.

The Laughing Kookaburra has a white chest and head, with a dark brown band running over its crown and through its eyes. Its lower back and tail are brown with black bars and its brown wings are flecked with blue. The Laughing Kookaburra is the largest member of the forest kingfishers, growing to 47 centimetres (19 inches) in body length.

Figure 8-1: The Laughing Kookaburra has a strong beak and band of brown feathers around its eyes.

The Blue-winged Kookaburra

To distinguish the Blue-winged Kookaburra from the Laughing Kookaburra, you only need to look at its head. Blue-winged Kookaburras lack the brown band that runs across the eyes. Other less obvious signs are a paler eye, a blue tail and a larger amount of blue in the wing of the Blue-winged Kookaburra — though that's probably obvious from its name! The Blue-winged Kookaburra is also slightly smaller than its laughing cousin — with a body size of around 40 centimetres (16 inches). And the cackle of the Blue-winged Kookaburra is also coarser and more abrupt than that of the Laughing Kookaburra. The Blue-winged Kookaburra is also sometimes called the Barking or Howling Jackass or Leach's Kookaburra. (What rotten nicknames for such gorgeous creatures!)

Blue-winged Kookaburras reside in the tropical and subtropical regions of Australia, in open woodlands, paperbark swamps, the trees along watercourses, and in canefields and farmlands from Shark Bay, Western Australia, across the Top End, through the Northern Territory to Toowoomba in southern Queensland. In the wetter tropical months, they rely on fish, crayfish, scorpions, spiders and earthworms for food.

Don't feed the birds

Kookaburras aren't usually dangerous until you try hand-feeding them. They have strong, large beaks that can injure your hand when they swoop down for the food you're offering.

Partners for life

Birdwatching enthusiasts say that Laughing Kookaburras pair for life. A bare nesting site is found — the hollow of a tree or a burrow dug out from a tree-dwelling termite nest — where the female lays 2 to 4 eggs. Males and females share in incubating the eggs and caring for the young. Older offspring help defend the nest and bring food to the parents and the newborn chicks. At about one month of age the young birds are ready to fly, and the parents and older offspring continue to protect them from predators.

Wearing a gardening or leather glove may protect you, but the best advice is, don't feed the birds! Also, kookaburras may attack you if you try to take their eggs or get too close to their young.

If you're injured by a bird's beak, wash and clean the wound with warm soapy water. Apply an antiseptic cream and clean bandage. Observe the wound daily for signs of infection (redness, swelling, weeping from the wound, pus and increased pain) and consult a doctor if an infection develops. A particularly bad gash of 3 centimetres (1 inch) or more may need stitching.

On the Wings of a Wedge-tailed Eagle

Unless you're a manic rockclimbing birdwatcher, you're unlikely to get close enough to be threatened by one of these magnificent birds. And you shouldn't be trying to! By simply observing them from a distance, you can admire the grace and beauty of their flight.

With a whopping wingspan of up to 2.5 metres (8 feet), the Wedge-tailed Eagle is Australia's largest living bird of prey, and one of the largest eagles in the world. Females are usually larger than males, by about a kilogram (2.2 pounds), with the average female weighing in at 4.2 kilograms (9.2 pounds) compared to 3.2 kilograms (7 pounds) for the average male.

Wedge-tailed Eagles are common throughout mainland Australia and Tasmania, from sea level to mountainous alpine regions, in wooded or forested land, and in open countryside. You may see them soaring high (up to 2,000 metres or 1.2 miles) on upturned wings, or perched atop trees or poles.

Group-hunting tactics

Working together as a group, Wedge-tailed Eagles can attack and kill animals as large as adult kangaroos. Up to 20 birds may be seen around a dead animal, although only 2 or 3 feed at any one time.

Adults are black-brown, though females are slightly paler than their mates. The Wedge-tailed Eagle has broad wings, a long, wedge-shaped tail, and legs that are feathered all the way to the base of their pale, powerful clawed feet (see Figure 8-2). Their intense, wide-ranging eyes and piercing stare allows them to locate even motionless prey, and their short, hooked cream-coloured beak is built to tear prey apart. The young are usually paler than the adults, with brown plumage and golden highlights.

Prey: Dead or alive

The Wedge-tailed Eagle feeds on a diet of rabbits, small marsupials, reptiles and birds, and can sometimes be seen on the side of roads feeding on carrion (dead animals). Live prey is captured during flight. Strong feet clutch and crush the prey, while deadly *talons* (claws) are driven through vital organs. A Wedge-tailed Eagle can lift up to 50 per cent of its own body weight. The eagle anchors its captured prey to its perch with its feet and then proceeds to pluck and shred it using its hooked beak. These huge eagles also store food on branches near their nest. Indigestible material, such as bones and fur, are regurgitated and expelled as pellets, landing below the perch. Wedge-tailed Eagles hunt solo, in pairs or in larger groups.

Figure 8-2: An adult Wedge-tailed Eagle has a wing span of up to 2.5 metres (8 feet).

Family business

Like kookaburras, Wedge-tailed Eagles pair for life. They build their nests in the highest location available, with a good view of the surrounding countryside, be it a tree, cliff face or even high bare ground. Their large nests of sticks, up to 1.8 metres across and 3 metres deep (6 feet by 10 feet), can be reused year after year by the same breeding pair. A breeding pair lives in the one area throughout the year, defending the area around the nest site from other Wedge-tailed Eagles — they have also been known to attack hang-gliders, abseillers, small aircraft and helicopters! Each pair also has a larger area, known as a home range, where they hunt for food but don't defend. These home ranges can overlap with the home ranges of other breeding and non-breeding Wedge-tailed Eagles.

Wedge-tailed Eagles are at their most dangerous during the breeding season, between April and September. You should never attempt to disturb an eagle's nest to steal eggs or to return chicks to the nest. Doing so greatly increases the risk that you'll receive a wound from an adult eagle's talons.

Because eagle wounds can be deep and dirty, with jagged edges, you need to take special care to clean and re-dress them daily. Seek medical attention if wounds are large (3 centimetres, or 1 inch, or more) or show any signs of infection — redness, swelling, weeping from the wound, pus or increased pain.

Feathered but Grounded

Australia is home to two families of large flightless birds: The Emu and the Southern Cassowary.

Emus live in all states of Australia except Tasmania (from which they were exterminated by early settlers). They prefer coastal grasslands, scrubland and wooded areas to graze upon, and don't live in rainforests and arid deserts. The Emu has been immortalised alongside the kangaroo on Australia's coat of arms.

Cassowaries are much less common, only inhabiting the dense rainforests of northern Queensland.

In the wild, both these birds are unlikely to attack you, unless you corner them or surprise a male incubating his nest. Then they can be very aggressive. Both creatures, particularly the Southern Cassowary, are capable of inflicting fatal injuries with their very large claws, or by delivering a swift and powerful kick.

'Old man' Emu

The Emu is the world's third-largest bird (behind the Ostrich and the Northern Cassowary), but is Australia's tallest, growing to 1.6–1.9 metres (about 5–6 feet) tall. An Emu can weigh between 30–45 kilograms (66–100 pounds) and is flightless, with only small redundant wings.

Adult Emus (see Figure 8-3) have long, shaggy, grey-brown feathers on their body, with short, downy feathers on their head, pointy beaks, and large aggressive-looking eyes. Their long necks have few feathers. Mature male Emus have blue streaks on their necks, whereas females and immature males have only grey-black skin. Emus have long, powerful legs with three forward-facing toes on each foot (no hind toes). Emu chicks (also shown in Figure 8-3) have soft cream and brown-striped plumage and spotted heads.

The name *Emu* comes from the Arabic word meaning 'large bird' and was first used by early European explorers who discovered Australia and some of its unique creatures.

The maternal male

Emus pair for breeding at about 20 months of age. The male builds a ground nest of grass, leaves and bark about 1 metre (3 feet) wide, then the female lays her clutch of 5 to 20 dark green eggs. After incubation begins, the male aggressively guards and incubates the nest. He sits on the nest for 55 days without drinking or feeding, living off the fat of his body. Chicks begin to leave the nest after 2 days, when they can begin feeding themselves. The young birds remain with the male for 4 to 6 months. During this time, their stripes fade and are replaced by dull brown feathers.

Figure 8-3: An Emu will attack to protect its chick.

Emus are nomadic, roaming over hundreds of kilometres in search of food and water. Grazing mainly on grasses, flowers and seeds, they also eat insects. They have few predators; lizards may eat their eggs, while Dingoes, foxes, and feral dogs and cats can attack Emu chicks, but few would-be predators can overtake or outrun an adult Emu, which can reach ground speeds of up to 50 kilometres per hour (30 miles per hour).

Emus can be confrontational — even in the wild they sometimes aggressively seek food from humans. They're big, have powerful claws, a large pointy beak and eyes that frighten most people. It's best not to mess with them; give up your food and retreat.

If you come face to face with an Emu:

- ✔ Keep an eye on the bird and back away slowly. Don't run.
- ✔ Protect your body with your backpack, picnic blanket or whatever you have at hand. Don't wave your arms or threaten to strike the bird.

Colourful cassowaries

Cassowaries have been described as the most dangerous birds in the world. The Australian or Southern Cassowary, shown in Figure 8-4, is Australia's second-largest bird, measuring 1.2–1.5 metres (4–5 feet) tall, and is one of only three species of cassowary found worldwide. The most recent recorded human fatality in Australia occurred in 1926 when a 16-year-old boy was kicked in the neck after provoking a Southern Cassowary.

The Southern Cassowary's head is devoid of feathers, but it has pale blue skin adorned with a horny helmet, and a neck covered with dark blue and purple skin, with long, swaying bright red wattles and a black shaggy body plumage. This makes for a very striking colour combination and works as amazingly good camouflage in the rainforests it inhabits.

The bird's long, stout and powerful legs have three claws each. The middle claw is the longest (120 millimetres or 5 inches) and sharpest, delivering the most harm in a fight. Females are usually larger and slightly brighter in colour than the males, weighing up to 85 kilograms (187 pounds), while the males weigh in at around 40 kilograms (88 pounds). Cassowaries, like Emus, are flightless birds with small wings.

Southern Cassowaries survive only in the tropical rainforests of far north Queensland, where they feed on fruits, fungi, snails and ground vegetation. They play a crucial role in dispersal of seeds for some native rainforest plants. The seeds in the fruits they eat travel straight through their digestive tract and are passed in their dung. This discourages other animals from eating the seeds, provides manure and scatters seeds throughout the rainforest.

Cassowaries can be very aggressive, especially males defending their young or if they themselves feel threatened.

Figure 8-4: A Southern Cassowary and its chick.

They use their feet to strike and tear into an intruder. Their claws can cause severe (even fatal) injuries. Cassowaries can also use their beaks to deliver powerful pecks, or their heads to butt a victim. Usually, attacks are provoked or result from the bird being frightened unexpectedly. Wounded or cornered birds are particularly dangerous.

Always observe these animals from the safety of your car, but if you come face to face with a Southern Cassowary:

- Don't feed it. The cassowary could turn on you and become very aggressive.

- Stand still, then slowly back away. Keep your eye on the bird. Don't run.

- As you retreat, protect the front of your body (use whatever you have at hand) or seek the shelter of a tree or some other structure. Don't strike out at the bird.

Report an attack by a cassowary to a ranger or the local authority.

First aid

Here's how to treat injuries sustained by an Emu or Southern Cassowary:

Wash the area with antiseptic solution or soap, or hydrogen peroxide (if available). Dry the wound and apply some antibiotic ointment before bandaging. Make sure you see a doctor if the wound appears to be infected (you note redness, swelling, weeping from the wound, pus or increased pain in the region). Also, consult a doctor if the wound is 3 centimetres (1 inch) or more, because it may need stitching.

If the laceration is major, stem the blood flow by applying pressure to the wound. Use a clean towel or shirt, or whatever you have at hand, and apply pressure for at least 15 minutes. If the injury is on a limb, raise the area above the heart to help slow blood flow, and call for an ambulance.

Going it alone

Cassowaries are solitary birds, only seen with a mate during breeding season, from May to November. After a short courtship, the pair mates; the female remaining only long enough to lay a clutch of eggs (2 to 5 large, blue-green eggs) onto a broad nest of leaf litter and grass on the forest floor. The female then deserts the nest (possibly to mate with several other males) leaving the male to incubate and care for the young until they can be independent (8–18 months). Chicks have cream and brown striped plumage when they hatch, which changes to brown during adolescence and black in adulthood.

Chapter 9

Australia's Living Icons

In This Chapter

▶ Avoiding the Platypus spur

▶ Escaping an echidna's spikes

▶ Sizing up kangaroos and the Koala

▶ Dealing with pesky possums

*A*ustralia has many unique and wondrous mammals. Some, like the Platypus, kangaroo and Koala, resemble no other animal on earth. These animals, along with the echidna and possum, provided naturalists who arrived with early Australian explorers with a lot of new material to discover. Today, scientists are still discovering new facts about these unique Australian icons.

In this chapter, we introduce some of these creatures' amazing characteristics and traits, including those that they use to defend themselves, which can be harmful to you if you get too close.

Because of Australia's unusually isolated position from the rest of the world, lots of Australian *mammals* — warm-blooded creatures that have mammary glands to feed milk to their young — are unique. For starters, many are *marsupials*. Of the world's 330 or so known species of marsupials, more than 200 of them are native to Australia and nearby islands. Marsupials (such as kangaroos, Koalas and possums) are creatures that give live birth to very immature, embryonic-like young, which then crawl into their mother's pouch to suckle and develop further. But even stranger than marsupials are the two monotreme species — the Platypus and echidna. These mammals actually lay eggs!

Monotremes also lack teeth — but they make up for this with other ways to defend themselves.

The name *monotreme* comes from two Greek words — 'mono' meaning 'one' and 'trema' meaning 'hole'. Monotremes differ from other mammals because they have only one body opening, called the *cloaca*, which acts as the reproductive tract (for laying eggs), urinary tract and anus.

One of a Kind: The Platypus

Platypuses live in streams, rivers and lakes, and are the world's most unusual mammal. Platypuses are monotremes, or egg-layers (as defined in this chapter's introduction), yet they also have mammary glands for feeding their young. They're covered in thick fur to help maintain body temperature, but have large, rubbery snouts (similar to a duck's bill), webbed feet — with a poisonous spur for defence on the hind legs in males — and broad flat tails. See Figure 9-1.

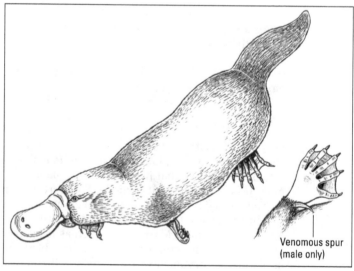

Venomous spur (male only)

Figure 9-1: The Platypus is sometimes called the Duck-billed Platypus. The male Platypus has a venomous spur on its hind legs.

The Platypus is so unusual that, when the first specimen was taken back to England by explorers in 1799, it was considered a hoax — a creature that a taxidermist had pieced together from other animal parts as a trick. But the Platypus is no joke, despite its strange appearance. The male Platypus spur carries enough venom to kill a large dog and to cause severe injury to humans.

A body like no other

A Platypus's body and tail is covered with three layers of brown fur. The inner layer traps air and keeps the Platypus warm; the middle layer provides a waterproof barrier; and the sensitive outer layer of longer hair helps the Platypus locate objects. The Platypus's fur is so dense and warm that these animals were hunted for their pelts until the early 20th century. The Platypus is now a protected species.

The body of the Platypus is 30–45 centimetres (12–18 inches) long; and the tail adds an extra 10–15 centimetres (4 to 6 inches) to its length. A Platypus can weigh up to 2.4 kilograms (5.3 pounds) and can live for up to 12 years in the wild. Males are usually a little larger than females.

The eyes and ears of a Platypus are located behind its bill, in a groove that closes while swimming underwater. Instead of using sight or hearing to locate prey during a dive, the Platypus uses its large, rubbery snout or bill to hunt for food and find its way around. It detects and locates its prey using *electro-receptors* in its bill (receptors sensitive to the movement of prey in the water, and to touch — see the sidebar 'Electro-reception finds the prey' for more), then stuffs the food into its cheek pouches before returning to the surface to eat. All four feet on the Platypus are webbed, but the front feet, which it uses more vigorously when swimming, have more extensive webbing than the hind feet. This webbing folds back when the Platypus walks on land.

The male Platypus has a venomous spur on each hind leg, where the foot joins its body (for more information, refer to Figure 9-1 and turn to 'Beware the spur' later in this section). Platypuses make a low growl when disturbed.

The shy type

The Platypus is found only in eastern Australia (including Tasmania). It constructs its nest at the end of long, deep burrows built into overhanging banks of rivers and lakes. The Platypus is a shy animal and only ventures out of its burrow to actively search for food at dawn and dusk. Its streamlined body enables the Platypus to dive and swim and fossick for food with ease. Platypuses feed on worms and insect larvae, and freshwater shrimps and *yabbies* (a type of freshwater crayfish), which they find in the mud of a riverbed with their bill or catch while swimming. During the daylight hours, the Platypus looks after its young, grooms itself and lies in the sun at the entrance to its burrow.

Beware the spur

The Platypus is one of the few venomous mammals on earth (and the single Australian venomous mammal). The male Platypus possesses venomous spurs, which he uses to defend himself against predators and during the breeding season against other aggressive males. The spurs can deliver a fatal injury to prey or rivals, and are capable of causing severe pain in humans. In the days when Platypuses were hunted for their pelts, dogs were sometimes mortally wounded when sent to retrieve a male Platypus from the water. People also get spurred when they incorrectly handle a Platypus. This can occur when trying to rescue a Platypus tangled in fishing line or an injured animal.

Electro-reception finds the prey

Monotremes — Platypuses and echidnas — are the only mammals known to locate food using *electro-receptors* to detect the electric fields generated by the movement of their prey. In the Platypus, electro-receptors and sensitive-touch receptors are located in the skin of its bill. These enable the Platypus to determine the direction and distance of any electrical and pressure impulses that are given off by moving prey. By sweeping its bill through the mud and water, like a mine sweeper, the Platypus can detect prey with speed and accuracy.

Platypus venom is not life-threatening to a healthy human. However, spurring is extremely painful — the spurs are deceptively sharp, some 15 millimetres (0.6 inches) long and can penetrate with force, deep into your skin. The venom causes spectacular swelling around the wound, which rapidly spreads over the affected limb, causing excruciating pain. The pain can persist for days or even months.

If you're spurred by a Platypus, stay calm and seek medical assistance as soon as possible. Don't apply a pressure immobilisation bandage. If bleeding is heavy, a firm bandage can be applied.

Don't use a cold pack or ice to try to reduce swelling, because this can intensify the pain and discomfort.

Spiky Surprise: Enter the Echidna

Two species of the spiky, insect-eating echidna exist in the world today. One is found only in the highlands of New Guinea, and the other in Australia and New Guinea. Along with the Platypus (covered earlier in this chapter) echidnas are the only monotremes in the world. The echidna also has a pouch to rear its young after hatching. The echidna's pouch faces backwards so that it doesn't fill with dirt when the animal is digging for food or to hide.

The Australian Short-beaked Echidna is found all over Australia, and can survive a range of temperatures and habitats — including the tropical lowlands of New Guinea. Echidnas are usually nocturnal (active at night), but in mild weather you may see one waddling along during the day. They have no fixed home, except when the female is suckling its young. Echidnas shelter under rocks, in the hollows of fallen trees or by burying themselves in the ground. Echidnas are sometimes mistaken for hedgehogs or porcupines because they're covered in sharp spines and have long snouts.

For most of the year echidnas are solitary animals, although each animal's territory is large and often overlaps with that of other echidnas. During the breeding season they use their fine sense of smell to locate one another.

A practical body

The echidna's stocky body (see Figure 9-2) is covered with hair and spines, and it has a short, stumpy tail and short legs, with sharp claws for digging. The coarse hair keeps the echidna warm and its long, sharp, hard spines (up to 5 centimetres or 2 inches long) protect the creature from predators. The colouring and length of the Australian Short-beaked Echidna's hair differs markedly across Australia, depending on the animal's habitat. For example, in the northern, warmer regions, the echidna's hair is light brown, but it becomes darker and thicker further south. In the colder climate of Tasmania, the echidna's hair is black. All Australian Short-beaked Echidnas have cream-coloured spines.

The Australian Short-beaked Echidna ranges in size from 30 to 45 centimetres (12 to 18 inches) and weighs between 2 and 5 kilograms (4.5 and 11 pounds). Echidnas in Tasmania grow larger than their Australian mainland cousins.

To grub for food, Australian Short-beaked Echidnas use their 8-centimetre (3-inch) pointy stiff snouts and extremely long sticky tongues. Their sticky tongues are perfect for catching ants and termites, or other insects and larvae, if they come by them. They make snuffling noises when they're searching for food, and use their front feet for digging or tearing open logs and termite mounds. Their hind feet point backwards and help push soil away from the body while burrowing.

Figure 9-2: An Australian Short-beaked Echidna.

Echidnas also use their back feet for grooming. Male echidnas have non-poisonous spurs attached to their hind legs.

Handle with care

When threatened, echidnas roll themselves into a ball or vigorously burrow themselves into the ground, displaying their sharp spines for protection. They don't attack.

In the wild, echidnas can live for 10 to 16 years. They have few natural predators. Goannas, Dingoes and foxes, along with feral cats and dogs, will eat young *puggles* (baby echidnas), but drought and fire are the main causes of adult death. Unfortunately, because echidnas move slowly, hundreds of echidnas are also killed or injured by cars every year.

The only way you're likely to be injured by an echidna is if you try to pick one up — then you're likely to be spiked. Its spines will penetrate your skin, which can cause local infections.

If you come across an injured echidna, or need to move one that's in danger, wear thick gardening gloves to protect yourself from its spikes. If you need to dig around the wounded animal in order to lift it, take a moment to determine which end is its head. Echidnas' snouts can easily be damaged or broken.

From egg to pouch potato

Pouch development in female echidnas marks the beginning of the breeding season (July or August). About three weeks after mating, the female echidna digs a burrow and lays one soft leathery egg directly into her pouch. Ten days later, the egg hatches and a blind, hairless puggle (the name given to these baby monotremes) emerges.

The puggle attaches itself to a milk patch inside the mother's pouch and suckles for the next 8 to 12 weeks. By the time the infant leaves the pouch, its spines have started to develop. The young echidna then begins to eat termites and ants, but remains close to its mother and continues to suckle for a further six months.

Living with the world's largest flea

Bradiopsylla echidnae, the world's largest flea, is 4 millimetres (0.15 inches) long, and lives on echidnas, along with ticks and other parasites. Echidnas have a long curved claw on each of their hind legs especially adapted to clean and groom between spines. They've also been known to go for a swim to help rid themselves of external parasites and to cool themselves in hot weather.

In fact, echidnas are good swimmers, swimming with only their snout and a few spines above the waterline. Despite this, they can still be infested with fleas and ticks, so if you do have to handle one, be sure to check yourself for parasites afterwards.

Lift the animal gently and place it in a sack or plastic bin, then take it to a vet. (Veterinarians are able to treat an injured echidna and have contacts for further care.)

Apart from the direct danger of being on the receiving end of an echidna's sharp spikes, echidnas are also usually infested with ticks and fleas, which can easily be transferred to you if you handle them, no matter how safely. To find out more on how dangerous ticks can be, refer to Chapter 6. For information on dealing with fleas, see Chapter 15.

If you get spiked by an echidna, thoroughly clean wounds with an antibacterial solution and hot water. Then dry and bandage the area. Check wounds daily for signs of infection (see Appendix A) and seek medical help if symptoms develop.

Australia's Ultimate Icons: Kangaroos and Koalas

Just about everyone agrees that kangaroos and the Koala are Australia's most iconic creatures. The planes of Australia's most well-known airline, Qantas, are adorned with the 'flying kangaroo' — its logo. And few tourists who visit Australia want to miss out on cuddling or touching a Koala.

But you need to approach both of these marsupials (pouched mammals) with caution, and avoid getting too close to them unless you're with a trained handler in a wildlife park, sanctuary or zoo (see Chapter 18 for more details).

Kangaroos: The mob's all here

Kangaroos are found throughout the Australian mainland, Tasmania and on some surrounding islands. In all, about 70 species of kangaroo exist, each suited to a specific environment, but the biggest and best-known kangaroo is the Red Kangaroo, shown in Figure 9-3, which in its normal standing position reaches 180 centimetres (almost 6 feet) in height. Other species, such as the Eastern Grey Kangaroo and the Western Grey Kangaroo, can grow to 160 centimetres (about 5.2 feet) tall. Female kangaroos are considerably smaller than their male equivalents. Wallabies, wallaroos, pademelons, potoroos, quokkas and tree kangaroos are smaller members of the kangaroo family.

Most kangaroos feed on grasses and leaves and live wherever water and their preferred food supply is available — which confines many of them to the east and the southwest regions of the country. Red Kangaroos need less water than Grey Kangaroos, and are more likely to be found in drier inland regions.

Boxing bucks and other group members

Bucks, boomers, does and joeys — each member in a kangaroo group has a special Australian name.

Kangaroos often travel in a group, which has its own unique name, called a *mob*. Male kangaroos in the group are called *bucks* or *boomers* — although the dominant, full-grown males are usually always referred to as the boomers. Female kangaroos are called *does*, and baby kangaroos are called *joeys* — a name they share with all other baby marsupials.

Kangaroo joeys are only about 2.5 centimetres (1 inch) long when they're born, then develop in their mother's pouch for the first six to nine months of their lives. Even when they're old enough to survive outside of their mother's pouch, they return to feed on milk for up to 17 months.

Kangaroos on the hop

Kangaroos have two powerful rear legs that allow them to hop forward at great speed for short periods — up to 70 kilometres per hour (43 mph) in the case of the big Red Kangaroo — and for longer periods at a comfortable speed of up to 25 kilometres per hour (16 mph). At low speed, perhaps when grazing, kangaroos move awkwardly, using their tails for support as they place their front paws on the ground and swing their hind legs forward like a pendulum.

Watch out for the right hook!

Although you can usually get within touching distance of a kangaroo in many wildlife parks and zoos, you're best to keep your hands off, unless you're with a professional handler. Attempting to feed a kangaroo — especially a large one — can lead to a sudden surprise attack. Although kangaroos are unlikely to attack without provocation, if you intrude too much on their space, they can stand up and push, punch, grab, tackle, kick, scratch, bite and deliver a direct attack at your head. And because a large kangaroo can weigh up to 90 kilograms (200 pounds) it can do a lot of damage — one killed a person, back in 1936. Kangaroos are also likely to attack you if you threaten their young.

If you receive blows or wounds from a kangaroo to your trunk, head or face, see a doctor to treat the wounds and to check for concussion. Also, ask for a tetanus shot if your immunisation is out to date.

Control a wound that's bleeding profusely by raising the injured part, and apply pressure to the wound with a clean cloth or even your fingers, if you have nothing else at hand, then seek medical help.

To treat minor cuts and scratches, wash the area with soap and water, and cover it with a dressing. Check daily for signs of infection (see Appendix A) and seek medical attention if symptoms develop.

Kangaroo collisions

Kangaroos are a real danger — to humans and themselves — when they're on or near roads between dawn and dusk, seeking food growing on the verges.

When startled by the noise of an approaching vehicle, or blinded by headlights, kangaroos often panic and hop into the path of vehicles. If this happens to you, hit the brakes and don't swerve. It's easier for the kangaroo to hop out of the way if your path is predictable.

Unfortunately, braking doesn't guarantee you'll avoid colliding with the kangaroo, and what happens next. In fact, some road fatalities have been attributed to collisions with kangaroos that have been hurled through windscreens. Other road fatalities have occurred when drivers have swerved to avoid hitting a kangaroo, and have either collided with another vehicle or lost control and left the road. Driving speed is usually the killer factor.

You're less likely to endanger yourself, the occupants in your vehicle, and the kangaroo, if you follow the wildlife road safety guidelines outlined here and in Chapter 2. Also, for information on what to do and who to call if you accidentally injure or kill a kangaroo, or you discover a kangaroo that someone else has injured, refer to Chapter 2.

Figure 9-3: The Red Kangaroo has two powerful hind legs and a tail strong enough to support its weight. Its colour can range from grey to red.

'Hey, do not disturb me' — the sleepy Koala

Koalas (see Figure 9-4) live in trees in southeast Australia, ranging from Queensland to South Australia (excluding Tasmania) — wherever they can find a suitable eucalypt forest. Their diet is very particular. Only about 12 species of eucalypt exist that Koalas will eat, and each Koala has its own preference or liking for one particular type of eucalypt tree's leaves within this group of 12.

Koalas are sometimes called 'koala bears', but they're most definitely not bears. Koalas are marsupials, or pouched mammals (as described at the beginning of the chapter). The name *Koala* is an Aboriginal word that means 'no drink'. And drink they don't — Koalas don't need a water source because they obtain all the water they need from the eucalypt leaves they eat.

Koalas are a protected species in Australia but are threatened by destruction of habitat, and by predators such as feral dogs and Dingoes, which grab them when they leave a tree and cross the ground to move to another. Like kangaroos and other wildlife, they can also become road kill. Also, a disease called *chlamydiosis* has affected many Koalas in Australia. The disease is treatable, but if it's not caught early enough it makes the Koala sterile.

Figure 9-4: A female Koala and her joey.

Koala kids

Like other marsupials, baby Koalas are called *joeys* and are born blind and hairless. The joey is less than 2.5 centimetres (1 inch) long when it crawls into its mother's backwards-facing pouch, where it stays for the next six months and feeds on milk from one of two teats. When the joey has outgrown the pouch, it rides on its mother's back, or rests against her chest. Before it begins eating eucalypt leaves, however, a joey samples its mother's droppings (faeces). This infuses its own gut with the necessary bacteria to help digest the eucalypt leaves that will be its diet for the rest of its life (about 15 years).

The long siesta

Even though Koalas live in *colonies* — in overlapping home ranges near other Koalas in a bushland region — Koalas are solitary animals that stake out their own territory. They spend their days asleep or dozing in the forks of trees, but just after sunset they begin moving around, feeding for a couple of hours, changing trees if necessary. It's during this brief period of activity that a male Koala may be heard 'barking' aggressively to warn other Koalas of its presence.

Koalas sleep for about 18 to 20 hours a day, but not because the fumes from the eucalyptus leaves make them drowsy or drunk. Koalas sleep so much to compensate for the extremely low-energy diet. Koalas consume as much as 600 to 800 grams (1.3 to 1.8 pounds) of eucalypt leaves each day, but these are very low in viable nutrients.

All fur and claws

Koalas have thick, light-grey fur and a white chest, fluffy ears, a broad flat nose and a short stumpy tail. They have strong limbs and paws equipped with long, sharp claws to help them climb trees. Each front paw has two thumbs and three fingers, giving them a powerful, well-balanced grip. The big toes on their hind paws are clawless, providing good friction for climbing and sitting. They also have two co-joined hind toes, which they use for grooming and removing ticks. Males are larger than females, and Koalas in the southern states tend to be larger than their northern relatives.

Koalas' sharp claws are dangerous. Koalas don't attack or lash out with their claws, but they instinctively use them to hang on or climb. Although Koalas look cuddly, if you mishandle an injured or frightened Koala, you're likely to get badly scratched. Also, never attempt to touch a Koala in the wild — in a few instances a frightened joey has mistakenly climbed a person instead of a tree, leaving behind some very nasty scratches.

If you're scratched by a Koala, thoroughly clean the wound with an antibacterial solution and hot water, then dry and bandage. If the scratch is a deep or large wound, consult a doctor, because it may require stitches.

Party On, Possums

'Hello Possums' is Dame Edna Everage's famous greeting. The phrase is an endearment because, to look at, possums appear very cute and cuddly. The trouble is, these inquisitive, big-eyed furry marsupials have become so used to people, they'll take food from you whenever they can, one way or another, even from your hand. They rob fruit trees, climb down chimneys to raid food inside buildings or steal from camp sites. They keep you awake at night as they scurry across your roof, going about their nocturnal activities, foraging and hunting for food. In short, possums can be downright pests. Because they carry ticks, they can infest you with ticks, too.

Living with the night-time raider

If you have a possum problem, you can contact your local council or authority to have them removed. However, many gardeners learn to live with possums, 'donating' the fruit at the top of a tree to the possums, while netting the bottom half. Possums like sweet-smelling mistletoe, and in this way they help protect native trees from this parasite. Still, some gardeners are often dismayed by the possum's fondness for their rosebushes.

Possums by the score

Australia is home to a large number of possum species, ranging from the small Western Pygmy Possum, weighing in at 13 grams (0.5 ounces), to species of large brushtail possums, tipping the scales at 10 kilograms (22 pounds). Possums are found in many habitats, ranging from rainforests in the north, to woodlands and heathland areas along Australia's coastline. Possums need fruit and flowers to eat, so they're not found in desert regions. Possums are common in camping grounds, suburban backyards and often nest next to the chimney in the ceilings of houses.

Some possum species, such as the family of gliders, have membranes stretching from their forefeet to their hind feet, enabling them to launch themselves from branches and glide from tree to tree. Other species, such as the Common Ringtail Possum, shown in Figure 9-5, have a tail that they wrap around branches, enabling them to grip and hang upside down to eat and groom.

No matter the size, these creatures have soft thick fur, a head with large forward-facing eyes, a small snout with long whiskers and large ears, and a tail longer than their body.

Living the high life

Possums are marsupials, carrying one or two joeys (baby marsupials) in their pouch. When the joeys become too large and emerge from the pouch, mother possums piggyback their babies, much like a Koala does, until they can fend for themselves.

Playing possum

In winter, when food is scarce, possums go into in a state of torpor — kind of like a state of hibernation. The possum curls into a tight ball, with its head tucked into its chest, then lowers its body temperature to near that of its surroundings, thereby conserving energy until warmer weather and food is available again.

During the day, possums rest in the hollows of trees or in fallen logs. By night, they emerge to feed on fruit, the nectar and pollen of flowers, and the buds and leaves of native and garden plants.

Possums are wary creatures, often sitting up a tree and surveying the lie of the land before proceeding to scavenge for food. Their quick jerky movements, coupled with their agility, make them hard creatures to catch.

Possums can be quite fierce when confronted, biting and scratching to protect themselves, their young or their territory.

 If you're injured by a possum, clean the wound thoroughly with soap (or an antibacterial solution) and very warm water to remove dirt, saliva and bacteria. Dry and bandage the injury. Check daily for signs of infection (see Appendix A) and go to the doctor if symptoms develop.

After a confrontation with a possum, check that you haven't picked up ticks. For more information on removing ticks or dealing with a larval tick infestation, refer to Chapter 6.

Figure 9-5: Two cheeky possum species: The Common Ringtail Possum (top) and the Common Brushtail Possum (bottom).

Part III
Bays and Beaches

Glenn Lumsden

*'The old 'fake shark fin on the back'
trick works every time!'*

In this part . . .

A trip to the beach . . . the prospect of a warm, sunny day, lapping waves and sea breezes — just sensational! Australia is a nation of beach lovers, but plenty of wildlife lives beneath the ocean waves and among the rock pools and reefs.

Although some of these are dangerous by reputation, we hope that these chapters entice you to enjoy the delights of the Australian coastline in safety. We provide you with ways to avoid becoming shark bait and tell you how to protect yourself from rock-pool nasties, such as the blue-ringed octopus. We also give you first aid advice on how to survive a deadly bite or sting should you be unfortunate enough to receive one.

Chapter 10

A Day at the Beach

*A*ustralians love the beach. Even in the cooler months you're likely to spot enthusiastic surfers braving the water. The Australian coastline — including the surrounding islands — stretches almost 60,000 kilometres (37,300 miles). The coast is lined with sandy beaches, rocky cliffs, shelves and mangroves. Many sandy beaches stretch for long distances. Some are very secluded or rugged.

A day at any Australian beach can be dangerous. If you don't wear sunscreen you're likely to get very sunburnt. And if you swim out of your depth in the surf, you're likely to drown. The shallow waters near the shore are also the breeding and feeding grounds for many different creatures, including some of Australia's deadliest — sharks and stinging jellyfish.

In this chapter, we take you to the beach and show you which creatures to avoid — no matter what — and those that you need to be wary of. We cover jellyfish that float or sail in on the currents, beach patrollers such as sea lions and some sharks, and fish that can bite, poison or give you an electric shock. We also cover noisy birds and biting flies, because even if you don't venture into the water, you may be attacked from the air.

Stingers in the Swim

You probably associate jelly (called *jello* in North America) with childhood parties and having fun. Yet jellies at the beach are just the opposite: Not fun! Some Australian jellyfish are deadly — their stings can be fatal within minutes. Their venom can stop your breathing and your heart from beating. If someone nearby doesn't recognise the signs of a jellyfish sting and give you mouth-to-mouth resuscitation and external heart massage immediately, you have little chance of survival.

Jellyfish vary in shape and size. Ranging from 10 millimetres to 2 metres (0.4 inches to 6.6 feet) across, they're composed of semi-transparent jelly, and move by way of ocean currents and simple rhythmic beatings of their bodies. They feed on plankton, small fish, worms and small *crustaceans* (aquatic creatures with hard outer skeletons, such as crabs, lobsters, shrimps, prawns and water fleas).

Almost all jellyfish have stinging cells, many will sting and leave itchy blotches on your skin. A few Australian jellyfish are mega-dangerous, though. In some cases, their sting is powerful enough to kill several adult humans, and even when washed up on the beach their stings can be life-threatening.

Deadly tentacles in the tropics: The Australian Box Jellyfish and Irukandji

The name *box jellyfish* describes a group of jellyfish with a four-sided, box-shaped body. Two species of box jellyfish stand out as the world's most deadly — the Australian Box Jellyfish (sometimes called the 'sea wasp') and the Irukandji.

Box jellyfish species, unlike other jellyfish that rely on ocean currents or the wind to get from one spot to the next, can propel themselves along at a speedy 2 kilometres per hour (1.2 mph), with rapid spurts of up to 9 kilometres per hour (6 mph).

The deadly Australian Box Jellyfish

The Australian Box Jellyfish has killed at least 70 people since 1900. An adult Australian Box Jellyfish contains enough venom to kill at least three men.

This extremely venomous jellyfish lives in the tropical coastal waters of Australia all year round (see Figure 10-1). However, this creature is mainly active during the wet season, from about November to April, especially after heavy rain. The Australian Box Jellyfish breeds and hangs around in coastal waters, estuaries and creeks. It prefers shallow waters and avoids the deep oceans and rough seas. You're unlikely to encounter the Australian Box Jellyfish near coral reefs, or in areas of seagrass.

The Australian Box Jellyfish (see photo in the colour section) is the largest of the box jellyfish. A fully grown adult has a cube-shaped body about 30 centimetres (1 foot) in diameter, and can weigh up to 2 kilograms (4.4 pounds). Its tentacles extend from the lower corners of the 'box' and grow up to 3 metres (10 feet) in length, and contract back to less than a metre after 'firing'. Each corner contains up to 16 tentacles.

Vicious venom

The stinging tentacles of the Australian Box Jellyfish contain venom that's very toxic. The venom

- Damages muscles and nerves, which stops your heart from beating and your diaphragm from moving (so you stop breathing too). The body collapses, so if you're still in the water, you'll soon drown.
- Attacks and damages red blood cells, which carry oxygen to all the cells of your body.
- Produces lesions and abrasions on the skin where the tentacles make contact, which tend to ulcerate (leave open wounds that are slow to heal), leading to permanent scarring and discoloration.

The tentacles stick tightly to the skin and may continue to release venom if not treated correctly, causing further harm to the victim. How much venom is released from a sting depends on how big the Australian Box Jellyfish is, how many tentacles have made contact with the body, the size of the victim and the sensitivity of the victim's skin. Children are affected very quickly.

All the better to see you with

Unlike other jellyfish, the Australian Box Jellyfish can see or at least detect and respond to light. It has clusters of 'eyes' on all four sides of its body. Some of these eyes possess corneas, lenses, irises and retinas. Some marine biologists believe this indicates that the Australian Box Jellyfish can actively hunt its prey.

As soon as possible, pour vinegar over the tentacles to deactivate them, soaking the area for at least 30 seconds. Only then can you attempt to remove the tentacles, very carefully, so that no more venom is released. Wipe the inactive tentacles off with a clean cloth or towel — don't rub the skin with sand, because this can scratch or abrade, adding to the victim's pain. *Note:* Never use methylated spirits, ammonia, urine or bicarbonate soda in place of vinegar, despite what you may have heard or read in the past.

In a serious case, when the victim loses consciousness, you need to perform immediate mouth-to-mouth resuscitation and external cardiac massage until the victim can be administered with antivenom. Call an ambulance and continue mouth-to-mouth resuscitation and external cardiac massage until help arrives, even if this takes longer than three minutes (this is because the body can shut down and the victim may appear dead and non-responsive, yet may still be alive).

See Appendix A for more information on mouth-to-mouth resuscitation and cardiac massage.

In mild cases, cold packs, painkillers and antihistamines can assist with pain relief and help ease swelling.

If you're swimming in areas where the Australian Box Jellyfish is found, make sure that you know the first aid procedures. And never swim alone in these areas — if you are stung, you may not be able to reach help.

Avoiding the Australian Box Jellyfish's tentacles

For many years, nobody knew what was causing swimmers in subtropical and tropical waters such excruciating pain or was sometimes killing them. This is because the Australian Box Jellyfish is semi-transparent and pale blue in colour, which makes it very difficult to see in the water. Today's better informed swimmers wear protective *rashies* or *stinger suits* — finely woven lycra swimming suits — or don't swim in tropical waters at all during the wet season.

In areas where these deadly jellyfish are found, popular beaches install *stinger nets* (netted areas) between November and April to keep them out. Some of these beaches may also be patrolled by lifesavers, who have vinegar and antivenom on hand.

Don't swim outside the nets or at unpatrolled beaches, especially between November and April.

The tiny Irukandji

Another deadly box jellyfish species is the Irukandji. This tiny, and we mean tiny, transparent jellyfish is only about 25 millimetres (1 inch) in diameter when fully grown, and is responsible for an average of more than 60 people being hospitalised each year. Two Irukandji fatalities have been recorded since 2002 — one at the Great Barrier Reef near Port Douglas, and the other at Hamilton Island in the Whitsundays, Queensland. Scientists believe that there may be at least seven different species of the Irukandji jellyfish, but only one has been formally identified and named (see 'What's with the name?' to find out more).

Although the Irukandji prefers deeper water, it's often swept into shallower waters by ocean currents.

Stinger nets can't protect you from the Irukandji. The Irukandji is small enough to get through the holes in the nets. The only way to protect yourself from its tentacles when swimming, snorkelling or diving in areas shown in Figure 10-1 — especially in Queensland, where most reported stings occur — is to wear a rashie or stinger suit, or to stay out of the water.

The Irukandji — see photo in colour section — has a single tentacle, up to 500 millimetres (20 inches) long, hanging from each of the four corners of its box-shaped body. The tentacles retract when prey is struck, in order to bring the victim to the Irukandji's mouth. Like the Australian Box Jellyfish, an Irukandji is very difficult to see in the water.

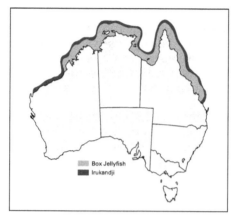

Figure 10-1: The distribution of the Australian Box Jellyfish and the Irukandji in Australia.

The sting of the Irukandji isn't always particularly painful — at first. Within half an hour or so of being stung, though, the victim develops Irukandji Syndrome: Lower back pain, then severe aches and shooting pains across the back and chest, muscle cramps in the arms and legs, and headache. Sweating can occur and the victim can feel anxious, nauseous and may even vomit.

In rare cases, water can accumulate in the lungs and the victim's heart begins to beat erratically, which can be fatal if not treated.

What's with the name?

The Irukandji is named after the Aboriginal tribe that once lived in the area around Cairns in north Queensland, where this stinger is most common.

The Irukandji jellyfish identified by scientists is called *Carukia barnes*, and was discovered by Dr Jack Barnes in 1964. He carried out a study in which he captured the tiny jellyfish and used it to sting himself, his son and a willing lifeguard to prove the jellyfish was dangerous to swimmers in coastal waters around Cairns, Queensland. For their trouble, all three were rushed to hospital with life-threatening symptoms, but they did recover. The jellyfish takes part of its scientific name from Dr Barnes.

Or the victim may suffer a brain haemorrhage. There's no antivenom for Irukandji jellyfish!

First aid consists of pain relief and reassurance. Vinegar can be used to reduce pain (alternatives to vinegar are currently under investigation), but call an ambulance if other symptoms occur. Hospitalisation may be required for further pain relief and heart monitoring.

More assorted jellyfish to avoid

Apart from the Australian Box Jellyfish and the Irukandji, no other jellyfish has been directly attributed with causing human fatalities in Australian waters. The stinging cells of all jellyfish can irritate, but several other species can cause very painful stings and potentially serious illness.

The Chiropsalmus jellyfish

The Chiropsalmus (see Figure 10-2), is a fairly small box jellyfish species that measures about 7 centimetres (3 inches) in diameter.

The Chiropsalmus is found in shallow coastal waters around the northeast tip of Queensland, as well as in the open sea in tropical regions, and is a powerful swimmer.

This jellyfish has bundles of tentacles up to 1.5 metres (5 feet) in length hanging from it body. Each bundle is made up of between 7 and 9 tentacles and has bands of pale purple stinging cells. Its tentacles are delicate and easily break off to adhere to a victim's skin or clothing.

Stinging tentacles

Most jellyfish have tentacles containing thousands of microscopic stinging cells. Each cell has a coiled stinging filament, as well as external barbs. When touched, the filament rapidly unwinds, launching into the target and injecting toxins. If the victim is food, it's dragged into the jellyfish's mouth; if the victim can't be eaten, the painful sting alone can still be lethal.

Little is known about Chiropsalmus venom, other than it's potentially lethal. A Chiropsalmus sting has caused deaths outside Australian waters. The effects of the venom are similar to, but less severe than, those of the much larger Australian Box Jellyfish.

Australian Box Jellyfish antivenom is sometimes used to counteract the venom of a Chiropsalmus.

Jimble

The Jimble, shown in Figure 10-2, is a small box jellyfish, ranging in size from 1.5 to 4 centimetres (0.6 to 1.6 inches) in diameter. The Jimble has a single tentacle hanging from each corner of its box-shaped body, and has stinging cells on both its body and tentacles.

The Jimble prefers cooler climates, and is found in southern waters from Coffs Harbour, New South Wales, to Albany, Western Australia. Stings can be quite painful, causing swelling and redness, which can remain visible for up two weeks. In severe cases blistering and scarring can result.

Lion's Mane jellyfish or Sea Blubber

The Lion's Mane (also called the Sea Blubber) has a flattened or squat bell-shaped body (see Figure 10-2). An adult has a bell measuring about 30 centimetres (12 inches) across, but specimens as large as 1 metre (3 feet) across have been found in cold waters. The Lion's Mane jellyfish is found in all Australian coastal waters.

Hanging from the bell of the Lion's Mane are eight clusters of tentacles. These tentacles can become detached from its body and still inflict their painful sting if encountered in the water.

Contact with tentacles may produce a burning feeling, which develops into severe pain lasting up to an hour. At times, sweating, muscles cramps, nausea and breathing difficulties may develop.

Mauve Stinger

The Mauve Stinger, shown in Figure 10-2, is a bell-shaped jellyfish that grows to about 12 centimetres (5 inches) in diameter. It lives in all coastal Australian waters. The upper surface of the bell is often covered in small wart-like lumps.

A thick bunch of tentacles hang from the centre of the Mauve Stinger's bell, with longer, thinner tentacles streaming from its edges. Body 'lumps' and tentacles contain stinging cells. Stings can be very painful and sometimes cause skin swelling and breathing difficulties.

Swimming when any kind of jellyfish is about puts you at risk of brushing against its stinging tentacles. The only way to minimise your risk of being stung — apart from staying out of the sea — is to wear a suitable rashie or stinger suit.

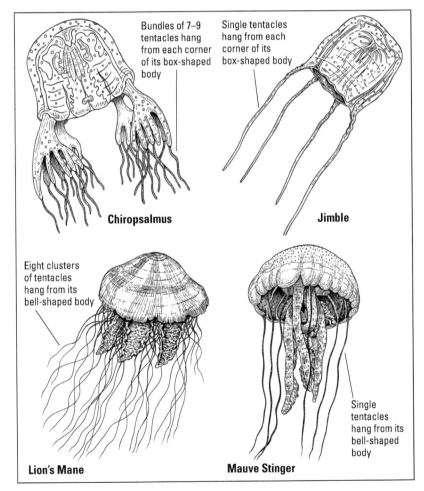

Bundles of 7–9 tentacles hang from each corner of its box-shaped body

Single tentacles hang from each corner of its box-shaped body

Chiropsalmus

Jimble

Eight clusters of tentacles hang from its bell-shaped body

Lion's Mane

Single tentacles hang from its bell-shaped body

Mauve Stinger

Figure 10-2: An assortment of stinging jellyfish.

Wearing a rashie, or a stinger suit, helps protect against stings because the jellyfish's tentacles can't grip onto the fabric. Before lycra suits became available, people living in tropical regions used to wear pantyhose to swim in the sea during the stinger season. Today, you can buy a suit that looks like a lightweight wetsuit, which covers most of your body (although you can still be stung where the suit doesn't cover your skin — around the feet, neck and head). Or you can don a full-body stinger suit that covers your whole body — top to toe — except for a small part of your face.

Wash Jimble and Chiropsalmus tentacles with vinegar before removing them carefully from the skin.

The jellyfish tentacles of the Lion's Mane and Mauve Stinger can be removed by washing the area with sea water and gently detaching them from the skin. Bathe the area in very warm water to reduce pain.

Painkillers (paracetamol and the like) are also useful in helping cope with the pain inflicted by stings.

Floating Bluebottles

The Bluebottle, also known as the Portuguese Man of War, is described as a jellyfish by most people — but it's not actually a jellyfish. A Bluebottle is a large, blue-coloured floating colony of differing *polyps* (soft jelly body parts), one of which forms a transparent gas-filled bladder that also acts like a sail on a yacht (see Figure 10-3). The bladder keeps the colony afloat. Below it hang three different types of polyps — some to catch prey, some to feed the colony and others to reproduce.

Sailing . . .

Half the sail on a Bluebottle is angled at 105° towards the wind and half at 105° against the wind. If the wind drives the Bluebottle towards the shore, one half of it is left stranded on the shore to die while the other half is swept back out to sea and survives. This venomous floating colony that makes up a Bluebottle is called a Man of War because it looks a little like an old Portuguese sailing ship. The Bluebottle is eaten by many marine animals, including sea turtles.

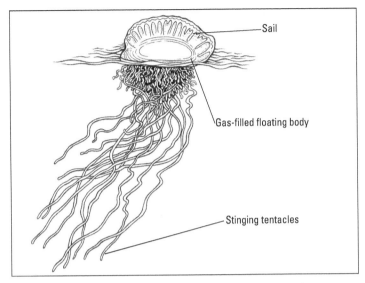

Sail

Gas-filled floating body

Stinging tentacles

Figure 10-3: The Bluebottle — a floating colony of venomous polyps.

A Bluebottle colony usually has a few very long tentacles — up to 30 metres (almost 100 feet) in length — and lots of smaller tentacles. These long tentacles have thousands of powerful stinging cells. When they make contact with prey, each stinging cell 'fires', pumping in a small dose of venom, stunning the victim. The stinging tentacles then contract, bringing the prey close to the shorter feeding polyps to digest.

Bluebottles are found in warm waters all around the world. Although Bluebottles have been responsible for deaths elsewhere, no fatalities have been recorded in Australia.

Bluebottles are carried by ocean currents along the east and west coasts of Australia, including Tasmania. In summer, prevailing winds can blow them ashore along the east coast.

On the beach, Bluebottles can break up and look like small, blue plastic bags. Children visiting the beach have been known to throw them at each other, receiving severe stings from the tentacles, which are still harmful after the Bluebottle has been washed up on the shore. And sometimes, Sea Lizards (covered next in this chapter), can be entangled in the Bluebottle's tentacles and can produce even more painful stings.

Bluebottle stings cause sharp, severe pain and typically leave a zigzag pattern of blotches on the skin. The pain can last for a couple of hours and can spread to joints and lymph glands. In some cases, the victim may suffer moderate back pain, and muscle cramps in the limbs, chest and abdomen. Bleeding can occur at sting sites and ulcers may form. This can lead to scarring later on.

When you're stung by a Bluebottle, usually only a few stinging cells have 'fired' — the rest fire as you try to brush the tentacles off your skin. The trick is to stay calm, and remove the tentacles with the pads of your fingers as soon as possible. Soak the affected area in very warm water to relieve pain and swelling. Painkillers (such as paracetamol) can also help with pain relief.

Sea Lizards

The Sea Lizard (also known as Glaucus) is an amazing little upside-down floating sea slug. This creature is blue and rarely grows larger than between 3 and 10 centimetres (1 and 4 inches) long, and yes, it's a slug, not a tiny reptile. The Sea Lizard just resembles a lizard . . . kind of.

The Sea Lizard (shown in Figure 10-4) has an air sac in its stomach that makes it float upside down. The creature's blue underside or 'foot' points skywards, and its silver-grey back points downwards, towards the ocean floor. Sea Lizards are most commonly found in warm, tropical waters, but can be sighted all around the Australian coast.

Figure 10-4: The Sea Lizard recycles jellyfish and Bluebottle stinging cells.

Sea Lizards are the ultimate in recycling. They can eat entire organisms — from small marine animals to the Bluebottle. After a meal, the Sea Lizard retains the most venomous stinging cells from its prey to use for its own self-defence. By concentrating these stinging cells in the tips of its tentacles, the Sea Lizard can deliver either a minor sting or one with a lot of punch!

Sea Lizards are carried to beaches by surface currents and winds. Often the first sign of their presence is the pain you receive from stings.

 Treat a Sea Lizard sting as you would a Bluebottle sting: Wash the area with sea water and gently detach its tentacles from your skin, using the pads of your fingers. Bathing the area in very warm water, and taking painkillers (paracetamol and the like) can help relieve the pain from stings.

The Lion of the Sea

An unprovoked attack on a 13-year-old girl at a beach in Western Australia in April 2007 is a reminder that any large creature — even those considered docile — can be dangerous. During this attack, the 200-kilogram (440-pound) sea lion leapt out of the water and bit the girl on the neck and face as she was being towed on a surfboard behind a boat. Her jaw was broken, three teeth were knocked out and a puncture wound to her neck missed a major artery by only a millimetre. The sea lion even made a second attempt to attack but was blocked by the boat.

The sea lion (shown in Figure 10-5) and seals are *pinnipeds*, meaning they have fin-like feet. Sea lions are longer and sleeker than seals and are able to use all four flippers to get around on land. Sea lions and fur seals belong to the family of eared seals. All other seals have no external ears — just tiny openings (or holes) through which they hear.

Sea lions are well known as intelligent, playful and social animals — their performances attract many visitors to marine parks like Sea World, on Australia's Gold Coast. An unprovoked attack on a human may seem out of character, but the 2007 attack was not the first: Although it was the worst, 17 other attacks have been recorded in Western Australia since 1978.

Figure 10-5: Sea lions are intelligent and social creatures.

No one really knows why sea lions attack. Experts say the 2007 attack may have occurred because the sea lion thought that the surfboard was its greatest predator, a Great White Shark, and it felt threatened, or that maybe the sea lion was a young male, over excited after competing with other males for a female to breed with.

Australian sea lions

Australian sea lion numbers are dwindling; scientists believe less than 12,000 live in the wild today. They're confined to the southern and southwest coasts of Western Australia and the coast of South Australia. The colonies that once existed further to the east of Australia were wiped out by seal hunters before the end of the 19th century. Australian sea lions are now a protected species. The Seal Bay Conservation Park on Kangaroo Island is home to about 700 Australian sea lions, living on the beach in their natural habitat, and are a popular tourist attraction.

Attacks by sea lions are rare and some — like the 2007 attack — may have simply been a case of being in the wrong place at the wrong time. You can avoid being attacked by a normally harmless sea lion or seal by observing them on land at a safe distance — of at least 10 metres. Don't forget that they can inflict serious injuries if you get too close.

Shark Attack

Shark attacks always attract headlines — even when they're not fatal. As a result, sharks have a reputation that they don't deserve.

Most sharks are harmless and aren't large enough to include humans in their diet or even risk getting close to them. They can be as small as the Pygmy Shark, which is only about 25 centimetres (10 inches) long. And even the largest shark of all — the Whale Shark (see Chapter 12), which can grow up to 18 metres (59 feet) long — is harmless unless you get in its way.

Worldwide, an average of about 50 shark attacks are recorded each year. Of those, about 10 are fatal. Australia has recorded 26 fatalities from shark attacks over the past 20 years. Survivors suffer injuries ranging from cuts, scratches and bruises, to amputated limbs.

The number of attacks per year is increasing, though. Some experts believe this is due to increasing human populations, over-fishing and depletion of the shark's normal food supplies, and the popularity of beach activities. Stories about sharks developing a taste for humans as prey are almost certainly false. Due to improvements in emergency services, medical procedures and public education, the percentage of attacks that are fatal has decreased from 60 per cent to about 20 per cent during the past 100 years.

Keeping track of shark attacks

The Australian Shark Attack File, based at Taronga Zoo in Sydney, compiles records of shark attacks in Australian waters since the first recorded attack in 1791. See Table 10-1.

Table 10-1	Australian Shark Attacks Since 1791		
State	*Total Attacks*	*Fatal Attacks*	*Most Recent Fatal Attack*
New South Wales	251	72	1993, Byron Bay
Queensland	231	71	2006, North Stradbroke Island
Victoria	35	7	1977, Mornington Peninsula
South Australia	49	19	2005, Glenelg Beach
Western Australia	85	13	2005, Houtman Abrolhos Island
Northern Territory	12	3	1938, Bathurst Island
Tasmania	21	5	1993, Tenth Island, Georgetown
Total*	684	190	

* As of November 2007 for all Australian states combined.

Pointing the finger at the aggressors

Of the approximately 170 species of shark that swim in Australian waters, only a handful have been identified in fatal unprovoked attacks on humans. These species are the Great White Shark (also known as the White Pointer), the Tiger Shark and the Bull Shark. All three often swim into shallow waters, near beaches. Of these, saying which is 'the most deadly' is difficult, because in many attacks the victim's confusion and fear prevents proper identification.

The Great White Shark

The Great White is the most famous of the deadly sharks, possibly because of its starring role in the movie *Jaws*. Growing up to 7 metres (23 feet) in length and weighing in at up to 3,300 kilograms (about 7,200 pounds), the Great White Shark feeds on a diet of fish, seals and sea lions — along with the occasional turtle, penguin or other sea creature. It prefers temperate waters, and in Australia can be found along the coasts of New South Wales, Victoria, Tasmania, South Australia and the southern part of Western Australia.

The Great White Shark is grey or bronze on top and white below. It has large, dark eyes and big white teeth — see photo in the colour section. The Great White Shark swims rapidly through the water, often attacking its prey by speeding in from a distance or from beneath. In many cases, attacks on humans may be instances of mistaken identity, because when the shark realises that its victim is not its usual prey, it usually swims away after a single bite. But with sharp triangular teeth up to 8 centimetres (3 inches) long, a single bite can be lethal.

The Tiger Shark

Growing up to 6 metres (20 feet) long, the Tiger Shark is potentially as dangerous as the Great White Shark. It feeds on fish, other marine creatures and just about anything it can swallow. Inhabiting shallow tropical and subtropical waters, Tiger Sharks have been found with dogs, sheep, bottles, cans and shoes, as well human remains, in their stomachs.

The Tiger Shark, shown in Figure 10-6, has dark eyes, a blunt head, a large mouth and very large serrated teeth. Young Tiger Sharks have distinctive dark stripes that fade as the shark grows older and larger. Attracted to disturbances in the water, Tiger Sharks often enter harbours and estuaries in search of food.

Figure 10-6: The Tiger Shark has a broad head, dark eyes and a large mouth.

The Bull Shark

The Bull Shark (shown in Figure 10-7) is a member of the family known as whaler sharks. This family also includes the Bronze Whaler, the Grey Whaler and the Blue Whaler. Whaler sharks are smaller than the Great White Shark and the Tiger Shark. The Bull Shark is the most likely to attack humans without provocation. It grows up to 3.5 metres (12 feet) in length.

This shark lives in tropical and subtropical waters as far south as Perth on the west coast, and Sydney on the east coast of Australia — very much the same waters as the Tiger Shark (covered earlier in this chapter). The Bull Shark has small, distinctive eyes and triangular teeth.

Figure 10-7: The Bull Shark is grey on top and white underneath.

The Bull Shark feeds on fish, including smaller sharks, dolphins and other marine creatures. Interestingly, the Bull Shark can live comfortably in freshwater as well as saltwater, so it swims into estuaries and rivers. The Bull Shark is believed to have been responsible for many attacks, including some deaths, especially on the Gold Coast, in the canals and in the surf.

Other sharks patrolling coastal waters

Apart from the Great White Shark, Tiger Shark and Bull Shark, several other sharks in Australian coastal waters are considered dangerous — and even deadly — mainly because of their size and weight. If provoked, they attack and can inflict serious or fatal wounds.

- ✔ **Bronze Whaler:** A bronze-coloured ocean shark, growing up to 3 metres (10 feet) long, which feeds on the sea floor in shallower waters. See Figure 10-8. Some have been found with human remains in their stomachs after being caught.

- ✔ **Wobbegongs:** A family of brown-coloured sharks up to 3 metres (10 feet) long, some of which are lightly spotted or banded. Wobbegongs have a flatter shape than most sharks, as shown in Figure 10-8. They can be found near the sea floor in coastal waters all around Australia.

- ✔ **Hammerheads:** A family of ocean sharks that sometimes venture into shallower waters. This shark has a distinctive hammer-shaped head (see Figure 10-8) and prefers tropical waters. Hammerheads can grow up to 5 metres (17 feet) in length.

- ✔ **Grey Nurse:** A grey-brown shark up to 3.6 metres (12 feet) in length. It lives in shallow coastal areas, often around underwater caves and ledges along Australia's eastern coastline from northern Queensland to New South Wales, and along the southwest coast of Western Australia. See Figure 10-8.

Protected sharks

You may be surprised to know that sharks, even the deadliest ones, need to be protected from humans. Sharks are more likely to be eaten by you — battered and called Flake in fish shops or maybe served as shark-fin soup at a restaurant — than you are likely to be eaten by them. About 20–30 million sharks are killed for food by humans every year. More are killed in the name of 'sport' and get caught up in shark nets or drum lines designed to protect swimmers and surfers from beach shark attacks. The killing of sharks in such large numbers, together with their slow rate of breeding, has put many species under threat of extinction. However, in Australia, some shark species are protected by law from injury or death at human hands. These species include the Great White Shark, the Grey Nurse and the giant Whale Shark.

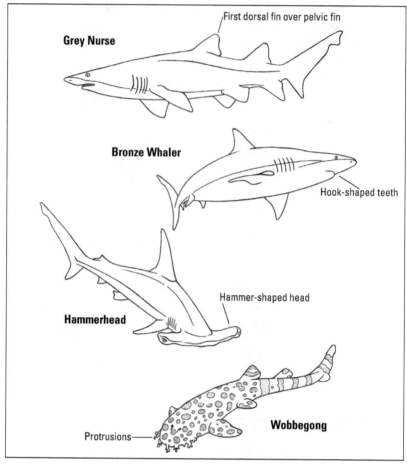

Figure 10-8: Other dangerous sharks in coastal waters.

The nature of an attack

Most unprovoked shark attacks occur in shallow waters close to the shore, where sharks are more likely to find large supplies of food.

The most common type of shark attack is known as the hit and run. This attack is usually the result of the shark incorrectly identifying a human as its normal prey, and often occurs in murky water or breaking surf. When the shark realises its error — usually after a single bite — it releases its victim and swims off. Most hit and run attacks are not fatal.

The less common, but more lethal, attacks occur when the shark deliberately targets a human as a source of food, or because the shark has been antagonised. These attacks are more likely to occur in deeper waters; they're vicious and frequently fatal:

✔ In a bump and bite attack, the shark circles and bumps the victim before striking

✔ In a sneak attack, the shark strikes without warning — it sneaks up on its victim

Remove the victim from the water. Apply pressure to the wound with a rolled-up towel to try and stem the bleeding. Raise the limb — reducing blood loss is critical. If the victim is wearing a wetsuit, don't try to remove it. Call for an ambulance immediately. Avoid moving the patient while waiting for the ambulance — movement increases blood loss.

Avoiding a shark attack

You can make the unlikely event of a shark attack even less likely by following these simple precautions:

✔ Avoid swimming in beaches that are known to be frequented by dangerous sharks. The Great Australian Bight — pardon the pun! — in southern Australia (refer to Figure 1-1) is especially well-known for visits by the Great White Shark.

✔ Take notice of warning signs and advice from local swimmers, surfers and beach lifeguards. Use patrolled beaches and swim between the flags. If a shark is spotted, leave the water quickly and calmly. Splashing about in a panic will attract the shark.

Lifeguards patrolling beaches around Australia keep a vigilant eye out for sharks and often raise an alarm (a loud horn) to ensure swimmers in the area hear the warning that a shark has been spotted, and to get out of the water.

✔ Never swim, surf or dive alone. Sharks are more likely to attack individuals. Avoid being stranded on your own, a long way from shore.

✔ Swim only in daylight. Sharks are more active at dusk, dawn and in the darkness of night.

✔ Keep away from large schools of fish, or seal or sea lion colonies, as well as river mouths, estuaries, deep channels and drop-offs. These are favourite feeding areas for sharks.

✔ Never swim near people who are fishing or where birds are diving for fish. These are signs that shark food is present . . . and perhaps also a shark.

✔ Never swim in murky water. You're more likely to be mistaken for prey and become shark bait. Also avoid areas where sewage enters the water because they attract schools of fish, which in turn attract sharks.

✔ Never swim with pets. Their erratic movements cause splashing that could attract a shark.

✔ Never go in the water if you're bleeding. Sharks can sense blood in the water over long distances and may soon arrive, looking for a feed.

✔ Avoid wearing shiny jewellery. It reflects light and looks like fish scales to a shark.

✔ Don't wear brightly coloured clothing — especially in murky waters. The contrast may make a passing shark curious enough to check you out. Even an uneven suntan might attract a shark. Similarly, if you wear a black wetsuit you could easily be mistaken for a seal.

✔ Scan the water for sharks before diving in from a pier or a boat.

Shark attacks of any type are rare. The risk of being attacked at the beach by a shark is very small. And the risk of a fatal attack is even smaller. You're more likely to be killed by a snake, a bee or a bolt of lightning — or even a falling coconut!

Fish to Steer Clear Of

Apart from some species of shark, most fish are completely harmless. But a few exceptions do exist and you may encounter some of these at the beach.

They're electrifying (Numbfish)

The Numbfish will shock you — literally! This fish gives you an electric shock that can be very dangerous if you have a heart condition. The Numbfish uses an electric shock to stun and numb its prey, and to deter predators. It has a large electric organ behind its head that can generate up to 200 volts, but a more usual shock is somewhere between 8 and 37 volts.

The Numbfish is found in Australian waters ranging from southern Queensland and south along the coastline, right through to the northwest coast of Western Australia (although it's rarely seen in Victorian and Tasmanian waters). They're usually seen singly on sandy and muddy bottoms in bays and estuaries, partially buried in the sediment. Numbfish live on a diet of crabs, worms and small fish, although some reports allege they also eat larger prey, including penguins.

The Numbfish is shaped like two discs, as shown in Figure 10-9, and has a short tail — it looks similar to a stingray (covered in Chapter 12). The Numbfish varies in colour, from grey or light brown to black. It has small eyes, which are raised when active. The body of the Numbfish is thick, blubbery and about 40 centimetres (1.3 feet) long, although specimens as big as 60 centimetres (2 feet) have been caught.

Figure 10-9: The Numbfish can provide a shocking experience.

A shock from a Numbfish feels similar to receiving a shock from an electric fence — temporary, delivering more than a little tingle, with no long-term effects. Normally, victims experience no other unhealthy side-effects. However, if you have a pre-existing heart condition, you may need medical assistance.

Poisonous puffers and porcupine fish

Pufferfish, sometimes called toadfish, are found in Australian waters from the tropics to the cool waters of the south, and have poisonous flesh. They live in the seaweed forests of coastal waters and estuaries. More than 35 different species of pufferfish exist, including the porcupine fish.

The porcupine fish has a more rounded body shape, but is otherwise very similar in appearance to its cousins. Australia has ten species of porcupine fish.

Pufferfish have a squat body, forward-facing eyes and a protruding snout, with powerful jaws and sharp front teeth, which they use to crush crabs.

The markings and colours of pufferfish, including porcupine fish, vary greatly, but they all have pale underbellies and spotted or striped upper regions to help camouflage them in their natural habitats. Both families of fish have spines on their stomachs and sometimes on their sides and back. The size, number and length of the spines varies from species to species.

A pufferfish poisoned Captain Cook

People have reported the dangers of eating pufferfish for many centuries. The earliest known records date back to the ancient Egyptians, then later the Old Testament book of Leviticus provided incentive by forbidding the eating of fish without scales. Later, Captain Cook reported his own near-death in New Caledonia (in the South Pacific) after eating a pufferfish in 1773.

Size varies from the small 9 centimetre (3.5 inch) Valentin's Sharpnose Pufferfish, to the 1 metre (3 foot) Giant Tropical Silver Toadfish. But most are between 15 and 25 centimetres (6 and 10 inches) in length.

Pufferfish can be ferocious predators, forming large hunting packs or schools, which go into a feeding frenzy in the presence of food. They have been known to bite divers and swimmers. Their teeth can penetrate a wetsuit!

All puffed up

A pufferfish will puff up right in front of your eyes if you disturb it (see Figure 10-10). When threatened, it puffs up by gulping in water or air. This causes its skin to stretch, revealing a spherical body covered in spikes, which can inflict a multitude of venom-loaded wounds to a predator.

Out of water, the puffing-up process creates a croaking sound, hence the nickname 'toadfish'. In some parts of Australia, pufferfish are also called 'blowfish' or 'blowies'.

The proof is on the plate

Pufferfish flesh is deadly to humans. The flesh and internal organs of the pufferfish contain a nerve poison, tetrodotoxin, for which no known antidote exists.

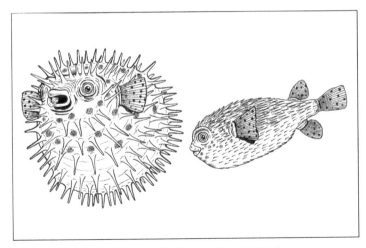

Figure 10-10: A pufferfish displaying its two different moods.

Poisoning is usually the result of eating incorrectly prepared puffer soup, *chiri*, or raw puffer flesh — *sashimi fugu*. In Japan, these dishes are considered delicacies and are only prepared by licensed chefs.

The appeal of eating pufferfish comes from being able to experience a mild feeling of intoxication, numbness to the lips and an all-over tingling sensation, caused by digesting low doses of the toxin. However, people die from high doses, whereas the after-effects of a lesser dose may result in headaches, vomiting, internal bleeding and paralysis. The element of risk in consuming pufferfish flesh adds to the excitement of eating this fish — apparently.

A victim of pufferfish poisoning may vomit and experience respiratory failure. Be prepared to support breathing by ensuring the airway is free and by providing mouth-to-mouth resuscitation. Perform cardiac massage (see Appendix A) if other life-sustaining systems fail, until an ambulance arrives.

You can also receive a nasty wound if you step on or handle a pufferfish. The spikes can penetrate deeply into your flesh. If this occurs, wash the wound thoroughly with an antiseptic solution and keep a watchful eye on it over the next few days in case an infection flares up.

Beware the Beach Birds

Most coastal sea birds pose no problem to beachgoers. But in spring and summer, when birds are establishing their territories, building nests and rearing their young, they can become aggressive towards other animals, including humans.

Training the birds, Hitchcock-style

The aggressive gulls and crows used in the Alfred Hitchcock thriller *The Birds* were trained to approach people for food scraps. Their squawking and chasing was encouraged so that when the actors appeared on set at these same feeding spots, this aggressive behaviour could be depicted in the film.

Pelicans, gulls and terns found near fishing spots and on popular beaches become accustomed to being fed by humans and can overcome their natural fears and learn to beg for food. If a human appears at a place where these birds are usually fed they may make an aggressive approach, hoping for a feed. This can look like an attack. Some birds get quite demanding in their attempts to gain food; squawking, hissing, begging, even pursuing a person.

Attacks can also occur if a bird is afraid or startled. Figure 10-11 shows some of the species most likely to attack at the beach:

- **Plovers:** The Spur-winged Plover (also known as the Masked Lapwing) and the Hooded Plover

- **Pelicans:** These marine birds can grow to 1.9 metres (6.2 feet) tall and have massive beaks

- **Gulls:** The Pacific Gull and the Silver Gull (or common seagull)

- **Fairy Terns:** Also called sea swallows, these species are threatened migratory seabirds

To avoid bird attacks when you visit the beach, look for signs of where birds may be nesting, and stay away from those places. In general, the closer you stay to the water's edge, the safer you'll be. (If you really want to feed the birds, ask a local where birds are usually fed or follow someone else's example.)

 Many beaches are valuable nesting sites and nurseries for sea birds, so it's important for birds to enjoy beaches and bays as much as you do!

Walking the dog

If you want to walk your dog on the beach, check the local by-laws covering the area you wish to visit. Laws can vary greatly from beach to beach. Some councils have beaches where dogs are allowed off-leash, others restrict hours and still others prohibit dogs completely. Be sure to keep your dog in check so that the bird life can feel safe.

Figure 10-11: Birds that may attack near water — a plover (top left), pelican (top right), silvergull (bottom left) and a tern (bottom right).

Buzzing Around (Flies that Bite)

Flies that bite are found worldwide and many carry diseases that affect humans and other animals. Australia is lucky, however, because none, except the mosquito (yes, the mosquito is a kind of fly!), is known to carry diseases harmful to humans. But flies are a nuisance and do cause painful bites, particularly in the summer months when they're in greater numbers and when you're likely to expose more skin.

Within Australia, the biting flies you're likely to encounter at the beach are shown in Figure 10-12. These include the March fly (or horse fly), the stable fly, the sand fly and the black fly. (Turn to Chapter 6 for information on mosquitoes.)

March flies and stable flies are vicious biters, seeking blood. They attack humans, livestock (such as horses and cattle) and pets. They're found throughout Australia in the warmer climates. March flies are a particular nuisance near water and on the beach in summer months.

Both the March and stable fly are solidly built, with piercing and sucking mouthparts designed to penetrate animal hides (so your socks or pantyhose are no protection against them). They're strong fliers, and are mainly active in summer, especially if the weather is still and sunny. They can grow up to 30 millimetres (1.2 inches) in length.

Sand flies are sometimes called biting midges, gnats or punkies, but no matter what you call them, they're often quite hard to see unless they're in a swarm — as often happens! Sand flies are tiny, at some 0.1–0.5 millimetres (less than 0.02 inches) long. They're the smallest of the bloodsucking flies, and only the female bites (the male feeds on nectar).

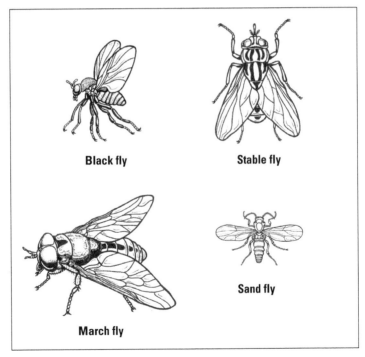

Black fly

Stable fly

March fly

Sand fly

Figure 10-12: Australia's annoying flies.

Sand flies are commonly found throughout Australia, around creeks and estuaries that open onto beaches, and their bites can drive you crazy.

Black flies are pesky rather than biters. Occasionally, swarms of these flies occur after floods in Queensland and in northwest New South Wales. In Australia, black flies aren't known to carry diseases, but bites can become infected if scratched.

You can avoid being bitten by flies if you use an insect repellent appropriate for your skin.

Fly bites can be itchy and quite painful, particularly sand fly bites. Symptoms can be relieved by using cold packs and applying antiseptic lotion or cream, or using products sold at pharmacies that specifically target and soothe itching. Scratching bites may lead to secondary infections. Also, some people experience allergic reactions to biting flies, although this is rare.

Chapter 11

Rock Pools, Reefs and Wrecks

* *

* *

*T*he creatures found in rock pools on the sea shore, or on rocky or coral reefs, are among the most beautiful in the world. In the rock pools formed by the rising and falling tides, you can find a variety of crabs, sea anenomes, sponges, sea snails, worms, barnacles and more.

Reefs — ridges or projections at or near the surface of the sea — can be built of coral, rocks, shipwrecks or even artificially. The nooks and crannies on reefs provide a range of conditions, each suited to different lifestyles, providing homes for a wide variety of marine life, from tiny colourful coral polyps to spiky fish and even whale sharks. The diversity is immense.

Exploring rock pools at the water's edge, or snorkelling or scuba diving around any type of reef, can provide hours of enjoyment and pure pleasure, as you encounter the wonders of marine life. In this chapter, we tag along as your friendly guides, providing information and commentary on the animals you're likely to meet.

Lethal Creatures in Rock Pools

Most rock-pool creatures are harmless to people. But some animals, like cone shells and blue-ringed octopuses, are capable of killing you while trying to defend themselves. Others can deliver a nasty painful bite, so you need to think twice before putting your hands into the shallow, clear water of a rock pool.

The deadly blue-ringed octopus

Unlike other, larger octopus species in Australian waters, the blue-ringed octopus is venomous. Normally, the little blue-ringed octopus is a rather dull creature; its brown body and banded tentacles provide camouflage while it hunts for food. When provoked, it quickly changes colour, becoming bright yellow, with distinctive bright blue rings (hence the name) appearing on its body and eight tentacles. This colour change serves as a warning signal to other living things that the creature is about to strike. (See photo in the colour section.)

The octopus hides out under rocks and in crevices, waiting for small unsuspecting crabs, shrimp or even wounded fish to venture by. Its prey, captured by its tentacles, is given a lethal bite of deadly venom, delivered by its sharp and powerful beak. Most blue-ringed octopuses are about the size of a golf ball and weigh about 90 grams (3 ounces).

A blue-ringed octopus may be small, but its venom is powerful enough to kill a human within minutes. In fact, one blue-ringed octopus carries enough deadly venom to kill about ten adult humans. Blue-ringed octopuses have been responsible for two fatalities in Australian waters.

At least four different species of the blue-ringed octopus live along the Australian coastline; some a little smaller than others. They're the most common of all Australian octopuses, living around shallow reefs and in rock pools, often washed up or trapped during tidal changes.

What every mother wants: Eight arms

The blue-ringed octopus has a nurturing side. Like other octopuses, it makes a den by piling up rocks to block up a hole or crevice. The den protects the octopus from predators (such as the moray eel) and provides a place to lay eggs and care for them. In autumn (March to May), males and females mate. The male dies afterwards, but the female blue-ringed octopus goes on to lay a clutch of about 50 eggs. She incubates them by holding them underneath her tentacles, close to her body, for up to six months. During this time she doesn't eat, and after her eggs hatch, she also dies. The new offspring mature and mate within a year.

Blue-ringed octopuses are not aggressive, but by the time you pick one up or step on one, and it emits its warning of electric-blue rings, it's too late! The bite might be painless, but the injection of its venom is life-threatening. A bite can be deep enough to penetrate a wetsuit.

The blue-ringed octopus injects a paralysing venom. Symptoms include nausea, loss of vision and speech, and an inability to swallow. Within a few minutes, paralysis sets in and you can have difficulty breathing. No antivenom is available.

Administer mouth-to-mouth resuscitation (see Appendix A) to keep the victim alive until he or she can be taken to hospital and placed on an artificial respiration machine. The venom gradually wears off after 24 hours, leaving no side-effects.

Cone shells: Beautiful but dangerous

Cone shells (or cone snails, as they're sometimes called) can be very beautiful: Ornate and colourful patterns adorn their snail-like shells. Because of this, people are inclined to pick them up, but this is a particularly dangerous thing to do because many cone shell species have venom powerful enough to kill humans. Children are especially at risk.

Some 80 different species of cone shell exist in Australian waters. Those that feed on marine worms have relatively weak toxin and pose no threat to humans; those that prefer fish in their diet have toxin strong enough to be fatal to people. Even specialists have trouble distinguishing the harmful cone shells from the harmless ones. The safest approach is to assume that all cone shells are dangerous.

Cone shells inhabit mostly mud and sand flats, rock pools and shallow reef waters, particularly at low tide lines. They're found along the northern, eastern and western coasts, but no further south than Sydney and Perth. Cone shells secure their food by 'spearing' their prey with poisonous harpoons! (See photo in the colour section.)

Cone shells have small, hollow, harpoon-laden teeth that are sharp enough to penetrate clothing. Each tooth has a reserve of deadly poison and a muscle, which shoots off a venom-filled barbed harpoon in the direction of any detected prey. Death occurs within seconds of an attack on marine animals. The deadly harpoon may be tethered and shot so that the cone shell can retrieve its kill after paralysis sets in. Or the harpoon may be detached so that the creature's venomous darts can be used in defence.

The harpoon mark made by a cone shell is triangular and is often quite small and deep. Sting symptoms vary, depending on the species of cone shell that's harpooned you. Most stings result in pain, swelling and numbness. Cone shell venom attacks the nervous system and can result in weakness, leading to a lack of coordination, blurred vision, loss of hearing and speech. Severe cases may result in death due to respiratory failure. One fatality has been recorded in Australian waters.

In the event of a cone shell sting, call an ambulance straight away. Apply a pressure immobilisation bandage (see Appendix A). If the victim becomes unconscious and stops breathing, use mouth-to-mouth resuscitation until help arrives.

Less Deadly Creatures Inhabiting Rock Pools and Reefs

Many fascinating creatures that live in rock pools or reefs are capable of giving you a very painful bite or sting. Although not all are lethal, many of these creatures, like crabs, urchins, starfish and worms, are best observed from a distance. And if you do become a victim of a bite or sting, it pays to know how to avoid a serious infection.

Cranky crabs

Crabs can look quite comical, with their sideways walk and wielding nippers. Many different kinds of crabs are found on Australia's rocky shores. Some like the shelter and quiet of a rock pool; others prefer rocky reefs and buffeting waves. Many use colour as camouflage, or decorate themselves with seaweed, or prefer to hide away in abandoned shells. Sponge crabs, for example, go as far as to carry live sponges or sea squirts on their backs to help them blend in.

Crabs range in size from that of a pea (pea crabs) to large hermit crabs averaging 80 millimetres (3 inches), to spider crabs with a claw span of 3 metres (10 feet).

The front pair of their five pairs of legs are nippers, with grasping claws on them to assist in feeding and defence. It's the nippers that can cause you harm. Crabs tend to hide from anything large, like you, but will swing their claws around and give you a nasty bite if you get too close.

Relatives of crabs — shrimps, prawns, crayfish and lobsters — can also give you nasty bites with their claws.

Many bites won't break the skin, but instead leave a bruise. But if a crab's nippers do pierce your skin, bathe the wound with an antiseptic solution and apply a dressing. Watch the wound over the next few days for signs of infection (see Appendix A).

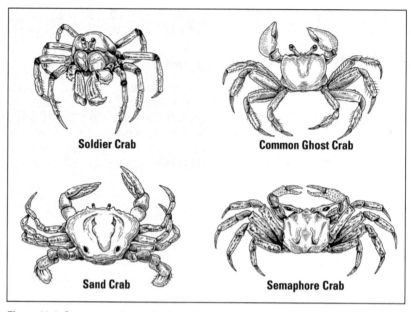

Figure 11-1: Common crab species found in rock pools.

Although crabs aren't venomous, the flesh of some crabs, especially those found in and around tropical reefs, is very toxic — toxic enough to kill humans. Several of these poisonous species of crabs are found in Australian waters.

Don't cook and eat a crab that you catch yourself unless you can reliably identify it as non-toxic. Australian mud crabs and the Sand Crab aren't toxic.

If you suspect food poisoning from a crab, seek urgent medical attention. Call an ambulance if the victim is unconscious.

Searchin' for urchins: Sea urchins

Sea urchins are found on rocky shores and reefs all around the Australian coastline. They're ball-shaped and easily recognised because of their many spines (see Figure 11-2).

The spines protect the sea urchin from predators (and curious humans). The sea urchin can move its spines, allowing it to get around.

The variety of sea urchins is amazing — they vary in colour and size. Not including their spines, sea urchins can range between 3 and 10 centimetres (1 and 4 inches) in diameter. The length and thickness of different species' spines vary considerably: Some are venomous and others are considered harmless. Most sea urchins feed by scraping algae from rocks using their rasp-shaped teeth, but some species cause coral destruction by grazing on polyps.

The spines on the Purple Sea Urchin are so strong, it can dig small holes into rocks. And the Needle-spined Sea Urchin has long thin spines up to 20 centimetres (8 inches) in length with barbs on the ends. These spines aren't venomous, but they can become deeply embedded in your foot or hand, providing a pathway to infection.

On the other hand, the Flowered Sea Urchin has rather short spines and, although it looks beautiful, it's exceedingly venomous. This species is believed to have delivered fatal stings to divers in Japanese waters, but no fatalities have been recorded in Australia. The most commonly sighted sea urchin along Australian shores is the Black Sea Urchin (see Figure 11-2), which can be found anywhere north of Perth or Sydney.

Although sea urchins look beautiful, we don't recommend that you touch them or pick them up. The sting of even non-venomous sea urchins is very painful and a wound can easily become infected. To avoid being stung while exploring rock pools, wear gloves and protect your feet with sturdy shoes.

Should you have a painful encounter with a sea urchin, carefully remove any spines protruding from the skin and bathe the wound in very warm water. Wash with an antiseptic solution and apply a bandaid. Check daily for possible infection (see Appendix A). See a doctor if you have any concerns.

Starfish

Sea stars or starfish, as they're commonly called, have flat, star-shaped bodies. Their size, and the number of arms they possess, varies greatly. Sea stars are able to regenerate injured or eaten arms, so you can find some weird-looking ones as well.

Most sea stars are harmless to humans and are a delight to observe. Their brightly coloured and patterned skins add to the beauty of any rock pool or coral reef. However, one sea star — the Crown of Thorns Starfish — is menacing and dangerous. It's dangerous to humans and a deadly menace on coral reefs. This sea star eats the living coral polyps on reefs, killing the very organisms that make up a coral reef.

The Crown of Thorns Starfish (see Figure 11-2) is found in rock pools and on coral reefs in the tropical waters of Australia. It's orange in colour and can have over 20 arms, and can grow to 70 centimetres (2.3 feet) in diameter. It also has venomous spines, so this is one big sea star you don't want to mess with.

If you accidentally touch or step on a Crown of Thorns Starfish, one of its long spines may pierce your skin allowing venom to enter your hand or foot, causing severe pain, which can last for several hours. Other symptoms include swelling around the puncture wound, nausea and vomiting. The swelling may last for several days. If you sustain multiple puncture wounds, the whole limb may be affected by the injury and you may suffer stiffness as well as swelling.

Figure 11-2: The Black Sea Urchin (left) and Crown of Thorns Starfish.

The Great Barrier Reef plague

In recent years, the Crown of Thorns Starfish has become a significant threat in some areas of the Great Barrier Reef, off the coast of Queensland, devouring huge numbers of coral polyps and leaving parts of the reef looking white, bleached and irreversibly damaged. The Crown of Thorns Starfish has few natural predators and if left unchecked can eat 6 square metres of reef per creature, per year.

To treat minor Crown of Thorns Starfish wounds, first pull out any spines that you can easily remove. Seek medical assistance to remove deeply embedded spines. Use very warm water to dull the pain and disperse the venom, then rinse and clean the wounds with sterile water. Apply some antibiotic ointment (if available) and a clean bandage.

Dress wounds daily and check for signs of infection (see Appendix A). Allergic reactions are possible, so seek immediate medical help if the victim displays rapid swelling and intense pain in a limb.

Bristle worms

Bristle worms or *polychaetes* are segmented marine worms covered with fine bristles or hairs (see Figure 11-3). A large number of different species fit this category: Rag worms, fire worms, blood worms, scaleworms, bait worms and sand worms, to name a few. Bristle worms range in length from 11 millimetres to 50 centimetres (0.4 to 20 inches) and have a pair of 'false' stumpy legs (which can be hard to see in some species) on each body segment to help them move around. They also come in different colours.

Bristle worms have diverse feeding habits. Some process muddy sediments for food; some graze on rocks for algae; others are successful predators. They're found in shallow rock pools, attached to reefs or at deep depths on ocean floors. They mostly live alone but a few species hang out with sea stars and other marine *invertebrates* (animals without backbones).

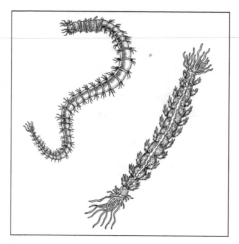

Figure 11-3: Bristle worms come in different shapes and sizes.

The larger species of bristle worm have jaws that can deliver a nasty bite, whereas fire worm can irritate your skin if you pick them up. Fire worms are fire-engine red in colour.

If you receive a bite or notice that your skin becomes inflamed or irritated after encountering a bristle worm, consult a doctor. You may have an allergic reaction or the start of an infection.

Reef Fish: Some Prickly Customers

The fish that swim in and around reefs or rocky shores are amazing — remember the movie *Finding Nemo*? Their diverse colour, shape and behaviour is incredible.

As well as being brightly coloured, many reef fish have spikes or spines on their bodies to protect themselves or to prevent predators from eating them. Some reef fish combine these features with the ability to make themselves look bigger, or to wedge themselves between rocks — making it hard for their attackers to dislodge them. Some creatures' spines are venomous, though, and some are just downright sharp and serrated, and can inflict really nasty gashes if you get too close. In this section we show you the creatures to be wary of.

Scorpionfish: Spiky swimmers

This group of spiky fish includes many dangerous and venomous species. The spines along the back and on the fins of scorpionfish can inflict painful wounds, which can also lead to infection.

The venomous Reef Stonefish

The Reef Stonefish (see photo in colour section) is reputedly the most venomous fish in the world. Camouflaged to look like a stone, this fish is mottled brown-green in colour, and often covered in algae and slime. Averaging about 35 centimetres (14 inches) long, Reef Stonefish have large fan-like fins behind the gills and a bulldog-like mouth. They live near the bottom of coral and rocky reefs, in shallow tropical waters, often hiding under rocks or ledges and sometimes burying themselves in sand or mud.

The fins just behind the gills at each side of a fish's head are called *pectoral fins*. The fins along the backbone are called *dorsal fins*.

Reef Stonefish are found in all Australian coastal waters north of the Queensland Gold Coast and Geraldton in Western Australia (refer to Figure 1-1).

Although no deaths from Reef Stonefish stings have been recorded in Australian waters, fatalities have been reported in other tropical regions. Stonefish antivenom is available.

The butterfly with a lethal sting

The Butterfly Cod, also known as the Common Lionfish, the Fire Cod and the Zebra Fish, is a very dangerous scorpionfish. Its brightly coloured striped body varies in colour to blend in with its surroundings. Some Butterfly Cod have fleshy tentacle-like protrusions above the eyes. They average a little more than 35 centimetres (14 inches) in length.

Found in shallow tropical and subtropical waters around rocks and coral reefs and estuaries, the Butterfly Cod (shown in Figure 11-4) is usually immobile during the day, hiding in unexposed places. It hunts for smaller fish, shrimps and crabs at night. The Butterfly Cod can be found in all Australian coastal waters north of Sydney and Perth.

Butterfly Cod have caused deaths in waters outside Australia. Their stings are extremely painful and their venom causes sweating, nausea, vomiting and, in severe cases, paralysis. Their stings are even dangerous hours after the fish die. Because these fish are favourites in home aquariums, stings are quite common.

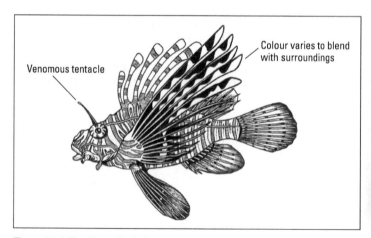

Colour varies to blend with surroundings

Venomous tentacle

Figure 11-4: The Butterfly Cod.

Spiky southern stingers: Goblins and Cobblers

Although the deadliest reef scorpionfish live in tropical and subtropical waters, a couple of dangerous ones are found in southern waters.

The Goblinfish (also known as the Saddlehead) and the South Australian Cobbler (also known as the Cobbler, the Scorpionfish and the Soldier Fish) can both be found along the southern part of the Australian coast between Perth in Western Australia, and central New South Wales. Both have venomous spines. Even minor stings from these fish produce intense pain. More severe stings can lead to sweating, vomiting, shock and even cardiac arrest. One death in Australian waters has been attributed to the South Australian Cobbler.

The Goblinfish (shown in Figure 11-5) has a deep groove behind its head and large circular eyes on each side of its face. This gives this fish a turtle-like appearance. Averaging about 20 centimetres (8 inches) in length, its colour varies from white to black. It rarely swims, preferring to move along the sandy sea floor in a hopping motion.

The Goblinfish favours protected coastal bays, seagrass beds and vegetated reefs in temperate waters. It hides under rocks and comes out at night to hunt crabs.

The South Australian Cobbler (see Figure 11-5) grows to about 23 centimetres (9 inches) long and has no scales. It has large brown splodges on its light-brown body, which provides excellent camouflage against seagrass and sand or mud.

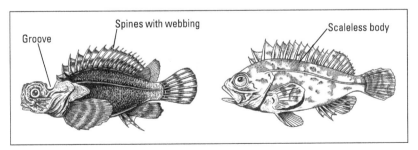

Figure 11-5: Two southern coastal scorpionfish: The Goblinfish (left) and the South Australian Cobbler (right).

The South Australian Cobbler is found inshore, close to beaches, in brackish subtropical waters along rocky reefs and in seagrass beds. Most active at night, the South Australian Cobbler lies motionless during the day. Smaller individuals feed mainly on shrimps and crabs, while the bigger ones eat fish.

Bullrout — a freshwater lover

Lurking in rivers, lakes and estuaries near the east coast of Australia is a spiny and slow-moving fish called the Bullrout. Like many of the spiny fish found in and around rocky sea shores, coral reefs and estuaries, this scorpionfish's sting delivers a forceful blow. Unlike other scorpionfish, however, you're more likely to find the Bullrout, shown in Figure 11-6, in freshwater rather than saltwater.

Bullrouts average about 18 centimetres (7 inches) in length, but can grow up to 35 centimetres (14 inches) long. They have 15 long and pointy venomous spines on their back and three more on their tail fin. These spines are used to defend themselves from predators.

Bullrouts prefer still water, where they can lie in wait for passing prey. Their dull yellowish-brown colour makes them very hard to see in muddy water, so they're very easy to stand on accidentally. And because they're very slow swimmers they can't get out of your way. Many stings occur when anglers remove this fish from nets or fishing lines.

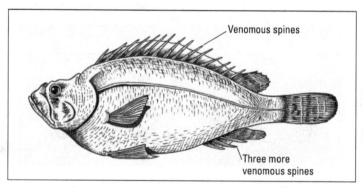

Figure 11-6: The Bullrout uses its spines to defend itself from predators.

If you're wading in rivers, creeks and lakes near the east coast, wear a sturdy pair of boots. When fishing in Bullrout territory, wear thick gloves when removing fish from nets or fishing lines.

At least 100 more scorpionfish to watch out for

Well, we don't have room to cover all types of scorpionfish, but assume that all 109 species of scorpionfish found in Australian waters are dangerous. Apart from the most dangerous ones covered earlier in this chapter, here are some others that you should be especially wary of:

- ✔ **Red Rock Cod:** Shown in Figure 11-7, the Red Rock Cod grows to 30–40 centimetres (12–16 inches) in length and has a light-grey to bright-red marbled pattern on its body, with small dark spots on its chest. It's commonly found on the rocky floor of coastal reefs in southern Australian waters.

- ✔ **Spotted Ghoul:** Averaging about 25 centimetres (10 inches) long, this fish has a brown-black freckled body and a lighter tan underbelly. It lives in tropical waters near the sandy floor around rocks, coral reefs and estuaries. See Figure 11-7.

- ✔ **Chinese Ghoul:** Also known as the Caledonian Stinger, the Chinese Ghoul is about 25 centimetres (10 inches) long and has a head covered with needle-like spikes. It's greyish brown in colour and lives on reef margins and in seagrass beds in the tropical north, often buried in sand.

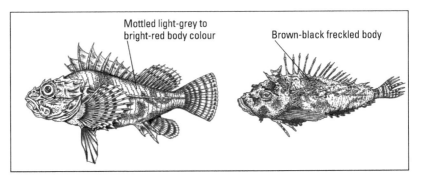

Mottled light-grey to bright-red body colour

Brown-black freckled body

Figure 11-7: The weird-looking Red Rock Cod (left) and the Spotted Ghoul (right).

Dealing with a scorpionfish sting

If you accidentally stand on, bump into or handle a
scorpionfish, you're likely to experience a very painful
injury, as one or more spines penetrate you skin and release
venom. The spines of some scorpionfish are sharp enough
to penetrate wetsuits. A sting can produce terrible pain,
which can last for several hours, and the venom can cause
headaches and vomiting. The severity of the symptoms
depends on the number of spines and the depth of
penetration.

Symptoms include:

 ✔ Rapid swelling of the affected body area, which may
   make movement of limbs very difficult

 ✔ Muscle weakness, leading to temporary paralysis,
   breathing difficulties and shock

 ✔ Death may occur in exceptional circumstances

First, immerse the affected area (most often a hand or foot)
in very warm water. This is thought to improve blood flow
and help disperse the venom. Seek medical advice and
try to get a good description of the species — this can be
important for treatment. Hospitalisation may be required.

As with all sting wounds, infection is possible, so closely
monitor a wound until it heals. The venom in the spines
remains active for days, so treat discarded spines with
caution.

Most people survive scorpionfish stings in spite of the great
pain. It may take several months for a full recovery, but if the
sting is left untreated, gangrene may develop.

Frog or fish?

The Bastard Stonefish is only one of several species of venomous Australian frogfish. A scaleless fish with frilly protrusions around the mouth and pectoral fins, this frogfish has a broad, flat head with forward-facing eyes — yes, it looks a bit like a frog or toad. Its mottled colours vary from grey to purple. The Bastard Stonefish averages 25 centimetres (10 inches) in length and has two venomous spines on its back, and three or four short, sharp spines on its gill coverings.

The Bastard Stonefish (see Figure 11-8) inhabits coral and rocky reefs all around Australia, except for the south coast and Tasmania. It sits motionless, waiting for its prey of prawns, crabs, shellfish and octopus to pass by. This fish can survive without water for some time and sometimes can be found on land, creeping along mudflats.

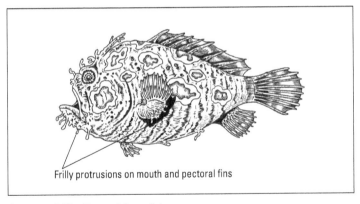

Frilly protrusions on mouth and pectoral fins

Figure 11-8: The Bastard Stonefish.

The stings from Bastard Stonefish spines cause severe local pain. Bathe the wound in very warm water to disperse the venom and reduce its effect. Pain can be relieved in most cases by taking generic painkillers; for more severe symptoms see a doctor. Check the wound daily for signs of infection (pus, increased redness, increased swelling) and seek medical help if symptoms occur.

Catfish: The fish with whiskers

Long whisker-like protrusions around the mouth give this species of fish its name. These 'whiskers' help the catfish (see Figure 11-9) to detect and locate its prey of shellfish, sea urchins and worms. All catfish can cause injury and inflict pain. They have a long, sharp, venomous spine embedded in a back fin and two others embedded in their chest fins that fire vertically. This poses a threat, particularly to fishermen, who catch them in their nets and have to untangle them.

Some 35 species of catfish live in Australian waters. They're found in rivers, estuaries and mud flats, and near rocky shores and coral reefs. They vary in size from 5 centimetres to 1 metre (2 inches to 3 feet) and are usually the same colour as their surroundings, so they're well camouflaged.

Although the catfish hasn't officially been responsible for any deaths in Australia, catfish venom causes severe pain for a number of hours, which can persist for up to 24 hours. A sting from an adult catfish can leave you feeling ill for several days.

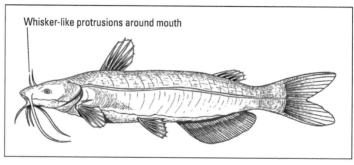

Whisker-like protrusions around mouth

Figure 11-9: A catfish has long whiskers.

FIRST AID Wash the injury in very warm water to disperse the venom. Use paracetamol or a similar painkiller to relieve the pain, but see a doctor if these drugs aren't helping. Check the wound daily for signs of infection and seek medical help if symptoms occur.

Old Wife or Zebrafish

Because this species of fish has black and white vertical stripes, the Old Wife is sometimes called the Zebrafish. You can easily identify them by their shape and colour pattern; they're usually seen in large schools of 50 or more.

The Old Wife, shown in Figure 11-10, is a cold-water species found in all Australian waters. The fish can grow up to 25 centimetres (10 inches) in length and inhabits sheltered regions provided by rocky reefs and seagrass beds. The Old Wife can also be found around piers and jetties. It feeds on marine worms and small shrimp.

Handling this fish poses a problem. The Old Wife has a number of fine, needle-like spines on its back that can inflict painful venom-loaded stings. Pain from a severe sting can spread from a wound site to an entire limb, and vomiting and headache can follow.

Figure 11-10: An Old Wife, with its distinctive zebra-like stripes.

Mumble, mumble . . .

When removed from the water, the Old Wife (or Zebrafish) makes a grumbling, grinding sound, like an old lady mumbling away to herself, so the name Old Wife stuck.

Bathe the injury in very warm water to reduce pain and the effects of the venom. Taking painkillers can also help to relieve pain, but if symptoms are severe see a doctor. Check the wound daily and seek medical help if infection occurs.

The nibblers: Rabbitfish

Rabbitfish nibble their food — algae — with protruding lips, rabbit fashion. Up to 16 rabbitfish species inhabit the Australian coastline, in all states but Victoria and Tasmania. Rabbitfish go by several other names — black trevally, black spinefoot and stinging bream.

Rabbitfish (see Figure 11-11) feed in schools around reefs, nibbling algae from the ocean floor or in and around seagrass beds. The seagrass-dwelling species are a dull greenish grey, but the coral reef species are brightly coloured — yellow and black. Rabbitfish range in size from 20 to 35 centimetres (8 to 14 inches), have flattened oval bodies with tiny scales, and a large number of spines along their back, with several more on their pelvic and anal fins.

Although the puncture wounds made by rabbitfish spikes are small, they inflict a painful venomous injury.

Treat puncture wounds by bathing in very warm water (to disperse the venom and help reduce pain). Take painkillers if required, but see a doctor if pain persists or if the injury shows signs of infection (see Appendix A).

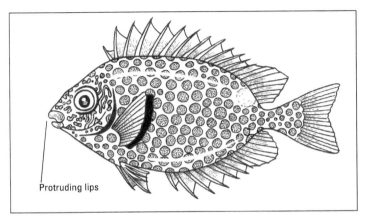

Figure 11-11: A rabbitfish nibbles food like a rabbit.

Quick on the trigger (triggerfish and leatherjackets)

Triggerfish and leatherjackets are similar in appearance (see Figure 11-12). They have a small delicate mouth with few teeth and small gill slits. They also have large raised spines on their head or back, and often possess a second spine or knob protruding from their chest. These spines can lock into position, making it difficult for predators to remove these fish from their rocky hiding places and making them tricky to eat.

The major difference between triggerfish and leatherjackets is the number of spines they have on their back: Triggerfish have three and leatherjackets have only one. None of these spines is venomous but in some species the spines are serrated, which can cause nasty, jagged wounds. Triggerfish also have large scales, whereas leatherjackets have tiny, hard-to-see scales.

Triggerfish tend to swim solo and feed on crabs, sea urchins and corals. Leatherjackets are usually seen in pairs or in small schools, nibbling on worms, sponges and seaweed.

Although triggerfish and leatherjacks have few teeth, some species can bite through a steel fishhook or a fisherman's finger.

Female triggerfish have been known to bite divers who stray too close to their nests. Most triggerfish and leatherjackets are aggressive towards other species of fish and use their raised spines to ward off would-be attackers.

Triggerfish and leatherjackets are found all around Australia, from the tip of Cape York, Queensland, to South Cape in Tasmania. The injuries caused by the spines of triggerfish and leatherjackets can be painful, but they lack venom, so you're unlikely to suffer any ill effects unless the wound becomes infected.

Keep the wound clean and dress it daily. Watch for any signs of infection (see Appendix A) and see a doctor if symptoms occur.

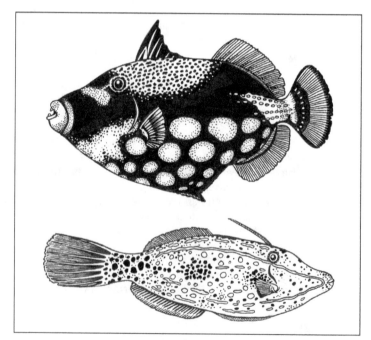

Figure 11-12: Common triggerfish and leatherjackets: The Clown Triggerfish (top) and Long-tailed Leatherjacket (bottom).

Surgeonfish on the cutting edge

Surgeonfish are so named because they have a sharp scalpel-like spine on each side of the base of their tail. When this fish is threatened or handled, these spines become erect and can inflict dangerous lacerations, cutting deeply into the flesh of an attacker. The pain associated with these injuries may last for several hours or even days; a fact that suggests that venom may be injected by the spines — scientists don't know for sure. Surgeonfish are also called Doctorfish or Spinetail.

Surgeonfish come in a range of sizes and colours, from yellow through to dark brown, with or without stripes or spots of blue or yellow. Most grow between 25 and 50 centimetres (10 and 20 inches) long. But the feature that best identifies them are those razor-sharp spines.

They live around coral and rocky reefs in tropical and temperate coastal waters from north of Perth, in Western Australia, and northern New South Wales on the east coast.

If you receive a deep cut from a Surgeonfish's scalpel-like spines, you may require medical attention to clean and stitch the wound. Deal with smaller cuts by cleaning with an antiseptic solution and bandaging. Pain can be relieved in most cases by taking generic painkillers; for more severe injuries, see a doctor. Check the wound daily for signs of infection, and seek medical help if pain persists or if the site becomes infected.

Sea Snakes: Life's a Beach

Sea snakes are a diverse group of reptiles — some 30 or more species inhabit Australian's coastal waters and estuaries. Most of these snakes prefer tropical and subtropical waters but some, particularly the Yellow-bellied Sea Snake, can be found as far south as Tasmania.

Although sea snakes tend to be found on or around rocky shelves and coral reefs, they're occasionally carried by currents or tides into the shallows of beaches, or into rock pools.

Sea snakes have elongated bodies, like land snakes, and most species have paddle-shaped tails as well, to aid movement through water (see Figure 11-13). They can stay underwater for long periods of time because they have an elongated lung that extends the length of most of their body, and fleshy flaps that close off their nostrils. They can also breathe through their skin. They shed their skin every two to six weeks. This process helps them to get rid of barnacles and sea parasites.

These aquatic reptiles can deliver a venomous bite when provoked. But most of the time they're quite harmless and curious creatures. Sea snakes only bite humans when they're handled or interfered with — either accidentally or deliberately.

Docile but toxic

Many species produce toxic venom to catch prey (small fish and eels) and for defence. But most bites experienced by humans are blanks — they're not loaded with venom, and aren't painful. Sea snakes are docile by nature and no human death due to a sea snake bite has been recorded in Australia, even though most species are highly venomous. However, one sea snake found in Australian waters, the Beaked Sea Snake (see Figure 11-13), has been responsible for many fatal bites in South-East Asia. Most bites have been sustained by fishermen emptying their nets and a few curious scuba divers.

When a toxic bite occurs, severe symptoms can be delayed for up to eight hours. Initial symptoms may include nausea, vomiting, difficulty in speaking and swallowing, impaired vision, and overall weakness, numbness or stiffness. More severe reactions include muscular pain, paralysis of respiratory muscles leading to breathing difficulties, and the production of dark-brown urine — an indication of kidney damage — then death. An antivenom effective against the bites of many sea snake species is available.

Don't cut or apply suction to the bite. Avoid movement by using a pressure immobilisation bandage around the limb (see Appendix A). At the first sign of noticeable symptoms, call an ambulance. Be prepared to give mouth-to-mouth resuscitation and heart massage if breathing difficulties follow.

Figure 11-13: The Yellow-bellied Sea Snake (top) and the Beaked Sea Snake (bottom).

Two family types

Two different families of sea snake exist; one that's exclusively aquatic and never leaves the water, and another type that spends its daylight hours on land (called a *sea krait*). Sea kraits return to the water between dusk and dawn to feed. On land, they remain within about 100 metres of the water, in grass and under rocks or logs, often in large communities of up to 100 or more.

All but two of the sea snakes species in Australian waters belong to the exclusively aquatic family. The two sea krait species are the White-lipped Sea Krait and the Black-lipped Sea Krait. Both species have extremely toxic venom. Sea kraits are restricted to tropical Queensland and northern New South Wales coastlines.

Sea snakes give birth to live, independent young, rather than laying eggs, like most terrestrial reptiles. This means they don't need to go ashore to reproduce.

The sea snake mating game

Like all snakes and lizards (and some earwigs), male sea snakes have two penises called *hemipenes*. However, only one is used for reproduction. During mating, the female pulls the male through the water by this penis. And she won't let go until mating is complete! The female is completely in control. Whenever she rises to the surface for a breath, the male — dragging behind her — also has to take a quick gulp of air.

Don't forget that sea snakes are venomous and must always be treated with extreme care. If you see one, leave it alone.

Coral, Glorious Coral

Australia's tropical and subtropical coral reefs are a maze of glorious colour, created by different living coral polyps. A *coral polyp* is a tiny fragile creature measuring about 3 millimetres (0.12 inches), which has a bony cup and tiny tentacles for feeding and protecting itself. Reef-building corals have tiny algae living in their tissues. Living together benefits both algae and coral, because algae gain shelter from the coral, while the coral polyp uses oxygen and food made by algae. The algae is responsible for the magnificent colours you see in corals — from bright reds and yellows, to different shades of blue, pale cream and pink.

Algae are plant-like organisms that usually live in water. Like plants, they photosynthesise — that is, they use light from the sun to produce their own food. The algae that live in the tissues of coral are microscopic.

Coral polyps can live individually, but many live alongside thousands of other polyps. By cementing together they form a strong and stable calcified structure or colony. Colony shape often gives rise to the name of the polyp that goes into making it — staghorn corals, horse's teeth coral, needle coral, brain coral, mushroom coral and soft corals (see Figure 11-14). Coral colonies with their strange shapes and formations grow together to form formidable reefs.

Figure 11-14: A coral reef is made up of a diverse range of corals.

Coral provides an abundance of food for marine life, and an assortment of nooks and crannies for vegetation to anchor to and animals to hide and breed in — an absolute visual delight for snorkellers, scuba divers and underwater photographers.

The bony skeleton of many coral colonies is hard and jagged, and can cause nasty cuts or grazes if stepped on, handled or brushed against. A wound can be embedded with tiny pieces of coral dust or animal or algal material. What might look like a minor injury can quickly develop into a nasty infected wound. Live corals have stinging cells, too, which can add to the severity of the wound.

If you receive a minor cut or graze, be vigilant in keeping the wound clean by washing it in an antiseptic solution and covering it with a fresh bandage each day. If the wound continues to sting after initial treatment, or is more than 3 millimetres thick, rinse it with vinegar to help dissolve and remove any coral dust, then rinse it a second time with clean water, and apply a clean bandage.

Check the wound each day for signs of infection (see Appendix A). Seek medical attention if symptoms occur.

The Great Barrier Reef

Australia's Great Barrier Reef is a World Heritage Area some 2,300 kilometres (1,400 miles) long — large enough to be visible from outer space. It stretches from just north of the southern Queensland town of Bundaberg up to and beyond Cape York. Really, though, it shouldn't be called a reef; it's comprised of almost 3,000 individual reefs. Over 350 different species of coral live there. Many of these same species can be found in the reefs off the Western Australian coastline as well.

Shipwrecks and Artificial Reefs

Shipwrecks and artificial reefs are located off the coast of every Australian state. They enable scuba divers to swim among a fantastic array of fish and other marine life. But along with these excursions into the deep, or not so deep, blue sea, comes the dangers that swimming in confined, out-of-the-way places can bring.

Shipwrecks — habitats and preserved history

Shipwrecks are underwater relics of preserved history; time capsules of the past. They can be very fragile. Inappropriate interference can lead to the destruction of not just an important historical site but also the collapse of a fragile haven for marine life.

Australia's shipwreck legislation ensures that wreck sites (and the marine life they support) are protected and preserved. Responsible diving benefits everyone's interests. Very few of Australia's shipwrecks require permits to dive. Local dive shops and specialist magazines can offer advice. Some popular dive sites are listed in Table 11-1.

Table 11-1	Australian Shipwreck Dive Sites
State	**Ship's Name and Location**
Queensland	*SS Yongala*, near Townsville; *SS St Paul*, off Brisbane
New South Wales	*SS Catterthun*, off Seal Rocks, central coast; *SS Tuggerah*, off Royal National Park in Sydney; and *SS Empire Gladstone*, off Merimbula on the south coast
Victoria	J-4 submarine and *SS Coogee* outside Port Phillip Heads; an assortment of vessels in Bass Strait, along the shipwreck coast, near Warnambool
South Australia	ex-*HMAS Hobart*, off the Fleurieu Peninsula
Tasmania	*SS Nord* and *SS Tasman* both near Eaglehawk Neck, Tasman Peninsula
Western Australia	ex-*HMAS Swan*, near Dunsborough

Artificial reefs

Artificial reefs in Australian waters have been constructed in the main to encourage fish to populate an area, or to provide a shelter for other marine species. In recent years, some reefs have been built specifically as *surf reefs*, to ensure suitable high-power waves for surfing. And yet other reefs have been built to combat beach erosion.

The materials used to construct these artificial reefs depends very much on the actual site — the type of sea floor, the wave and tide action, the depth and, of course, the purpose for which the reef is being built. Materials with holes and crevices are great for encouraging fish, but terrible for encouraging a long wave. All materials are free of pollutants.

The dangers within

Whether we're talking about a shipwreck or artificial reef, there are common dangers beyond those that are present in natural rocky shelves and coral reefs. Sunken boats and ships, cars and concrete pipes pose many problems for divers. By law, any designated diving site must have hatch lids, sharp beams, ropes and cables removed to ensure divers' safety. In addition, a site is inspected to ensure two exits exist in every confined space. This ensures that when divers come face to face with a creature they've disturbed, like a shark, at least one escape route is available — for either the diver or the shark!

What you see when making a dive on a shipwreck or an artificial reef designed to support marine life will vary, depending on the location — in tropical, subtropical or temperate waters. But you need to be on the alert to avoid accidental encounters with scorpionfish, other spiky fish and sea snakes sheltering inside. Although not aggressive, if they're frightened their natural defence mechanisms may come into play.

Observing, not touching, and following the scuba-diving rules of never diving alone and always returning to the surface well before your air tank is due to run out, is the safest way to investigate underwater life.

A history lesson

You may be thinking that the construction of artificial reefs is a relatively new idea, but the Ancient Persians constructed an artificial reef or barrier to the entrance to the Tigris River to prevent pirates landing. And around 264 BC the Romans built a reef to trap the enemy ships within the Carthaginian harbour in Sicily.

Japan has used artificial reefs to grow and harvest kelp since the 1600s. And in the United States, wooden artificial reefs were built in the 1830s to provide fish habitats.

But the idea of constructing artificial reefs to improve surfing is fairly recent. Australia's first man-made surf reef is at Cottesloe Beach, Perth, Western Australia. This reef is constructed of large granite rocks placed in a pyramidal shape to form an appropriate breaking wave form that suits surfers.

Chapter 12

Dangerous Creatures of the Deep

*W*hen you venture beyond the shoreline and shallow reefs into deeper waters, you enter a whole new world of fascinating creatures. In this chapter, we cover some striking marine animals — sharks, whales, stingrays and spiky deep-water dwellers. Some of these creatures are unlikely to attack you but can still inflict painful injuries; others are potentially harmful but are rarely encountered. Some are deadly, though, so you need to know which ones to avoid, especially if you're a diver or deep-sea angler.

Sharks in the Deep

Sharks are at their most dangerous when they're feeding — and most sharks need to visit shallow coastal waters and reefs for a meal. But some sharks only live in the deep ocean. You're most likely to encounter them if you're in a boat or unfortunate enough to be stranded in the water in the middle of the ocean.

Watch out for the White Tip

Although the Oceanic Whitetip Shark ranks among the four most dangerous sharks in the world, you're not likely to encounter one while swimming or surfing at the beach. Unlike the Great White Shark, the Tiger Shark and the Bull Shark, the Oceanic Whitetip Shark, shown in Figure 12-1, is a deep-ocean shark, mostly encountered by game-fish anglers or divers.

Adult Oceanic Whitetip Sharks are usually less than 3 metres (10 feet) long, but can reach a length of 4 metres (13 feet). Their sharp teeth are perfectly shaped to tear the flesh from their prey. They prefer water temperatures greater than 20° Celsius (68° F), so they're usually found in tropical and subtropical waters. But these sharks sometimes venture as far south as Tasmania.

Figure 12-1: The Oceanic Whitetip Shark has a blunt nose and a white-tipped rounded fin.

In the open ocean, the population density of fish is much less than around reefs and in the shallows. This means that the Oceanic Whitetip has to swim long distances to find food, so any smaller creature in the water (including humans) is on the menu. Any diver in the water, or survivor of a shipwreck or plane crash in the ocean, is in danger. Oceanic Whitetip Sharks are believed to have killed numerous men who initially survived the sinking of ships during World War I and World War II.

The speed king of sharks

The Short Fin Mako (see Figure 12-2) is the fastest shark of all. Its streamlined shape is perfect for high-speed swimming. The Short Fin Mako is also known as the Blue Pointer because it has a blue-shaded upper body and pointed snout. It lives in both tropical and temperate waters, and can be found off most parts of the Australian coast.

The Short Fin Mako is believed to have been responsible for several attacks on humans in Australian waters. The most likely victims are game anglers fishing for sharks and other large fish. When hooked, the Mako's great speed allows it to soar spectacularly into the air — occasionally landing in a boat and attacking those on board.

Figure 12-2: The Short Fin Mako grows up to 4 metres (13 feet) long.

Dealing with an attack at sea

Sharks can inflict terrible wounds (refer to Chapter 10 to find what they're capable of). In the event of an attack, remove the victim from the water and apply pressure to the bite wound with a rolled-up towel to stem the bleeding, and raise the limb. Reducing blood loss is critical. If the victim is wearing a wetsuit, don't try to remove it. Call for an ambulance immediately. Avoid moving the patient while waiting for the ambulance — movement will increase blood loss.

True Giants of the Sea

It sure is nice to know that the biggest creatures in the ocean aren't the most dangerous. In this section we look at two different types of marine animals: The Whale Shark, officially a shark but similar in size and some of its behaviours to a whale, hence it's name, and a few species of whale that inhabit Australian waters.

A whale of a shark

The largest shark of all, the Whale Shark, isn't usually harmful to people. However, any animal measuring up to 18 metres (about 60 feet) long and weighing as much as 15 tonnes (33,000 pounds) can be dangerous if you get in its way. It could capsize your boat, leaving you vulnerable to more dangerous sharks, or to currents that could drag you under. The Whale Shark (shown in Figure 12-3) won't bite you, though. It has about 3,000 tiny hooked teeth that are only a few millimetres long, which can't bite through or tear flesh. Like whales, these giants feed on *plankton* and tiny fish, crabs and squid. They filter these small animals from the water through their gills.

Plankton is the name given to a variety of animals and plants that drift with currents in the water. There are billions of these organisms in the ocean. Algae, jellyfish and the larvae of sea snails are some examples of plankton.

Swimming with the giant shark

At Ningaloo Reef, Western Australia, you can swim and dive with Whale Sharks. They converge on the reef from late March until July every year for a huge feed. Strict rules prevent swimmers and divers from interfering with the Whale Sharks while they feed. No boats are permitted within 30 metres (33 yards) of a Whale Shark and swimmers and divers must remain more than 3 metres (10 feet) from the shark's body and 4 metres (13 feet) from its tail.

Broad flat head with enormous mouth

Figure 12-3: A typical Whale Shark is up to 18 metres (60 feet) long.

Whale Sharks have light-coloured stripes and spots on their bodies to help them blend in with their surroundings. They live in deep water in the open sea but also visit the coast, swimming long distances to find large supplies of plankton — usually in warmer waters, in subtropical and tropical regions of the Australian coast. They're rarely sighted except when feeding near the coast — in fact, their existence wasn't even known until 1828. Since then, Whale Sharks have been hunted and killed in large numbers for their meat. Over-fishing has seriously threatened their survival. Today, though, they're a protected species in Australia.

Whales: Just cruising by

No whale — not even the gigantic Blue Whale — has been known to deliberately attack humans. But you still need to observe these magnificent creatures from a distance — mostly to protect them from stress but also to ensure that they don't accidentally harm you or the boat you're in.

Whale watching, when these creatures are migrating up or down the Australian coastline, is big business. Strict guidelines are in place to stop tourist operators approaching whales too closely in boats. Whale-watching boats are required to stay at least 100 metres (110 yards) away from each side of any whale and aren't permitted to wait in front of a whale or approach from behind.

Demystifying the Killer Whale

You'd expect a large creature with the name Killer Whale to be dangerous — even lethal. But they're not true whales. Killer Whales are the largest members of the dolphin family, reaching lengths of up to 10 metres (33 feet). They're certainly killers. Killer Whales kill and eat other marine creatures, including fish, seals, sea lions and even Great White Sharks; however, no human fatalities have been recorded in Australia. The only attacks that have caused injuries to people have occurred in marine parks, where captive Killer Whales have forced their handlers underwater and attacked them.

The largest creature on earth, the Blue Whale, is almost twice as long as the Whale Shark, growing up to 33 metres (108 feet) long. Blue Whales are found in Australian waters but they're rarely sighted because they prefer to remain in colder waters a long way from the coast.

The whales you're most likely to spot off the coast of Australia are the Southern Right Whale (up to 18 metres or about 60 feet long) and the Humpback Whale (up to 19 metres or about 62 feet long). These giants of the sea can be observed migrating north from their feeding areas near Antarctica to warmer waters to breed between May and September, and south between September and November.

- ✔ Female Southern Right Whales and their calves can be seen all along the south coast of Australia and around Tasmania. The most frequent sightings are along the southwest coast of Victoria, the Great Australian Bight in South Australia, Cape Leeuwin on the southwest tip of Western Australia and in Storm Bay in Tasmania.

- ✔ Humpback Whales are seen all along the east and west coast. Favourite viewing locations include Hervey Bay in Queensland, and Byron Bay, Port Stephens and Jervis Bay in New South Wales.

Stingrays: The Graceful Movers

Stingrays are graceful creatures (see Figure 12-4). Their flat bodies are perfectly shaped for gliding through the water and burying themselves in the sand on the sea floor. They feed on shellfish, crabs, fish and other small sea creatures.

Numerous stingray species live in Australian waters. Their topside colours vary, but all are a very pale colour underneath. Many of them are named after their colour or shape, such as the

- ✔ Black Stingray
- ✔ Southern Eagle Ray (also known as the Bat Ray)
- ✔ Eastern Shovelnose Ray
- ✔ Blue-spotted Stingray

Figure 12-4: A stingray has a round, flat body and a long, tapered tail.

The largest stingray in Australian waters is the Smooth Stingray, which lacks the small tooth-like structures on the skin (called *denticles*) that other stingrays possess. The Smooth Stingray can grow to more than 4 metres (13 feet) long, and weighs around 350 kilograms (770 pounds). It can be found in coastal waters in the southern half of Australia at depths of up to 170 metres (186 yards).

Stingrays and sharks don't have bony skeletons like other fish. Their skeletons consist of cartilage, which is flexible and elastic. It's similar to the cartilage that makes up the external part of the human ear.

On the defence

Most stingrays have long, tapered tails, with one or more barbed stinging spines. Stingrays aren't aggressive animals. Their stinging spines are only used in defence. But when these stingrays are cornered or threatened by a predator, they can react by driving their barbed spine into the antagonist. Venom is injected from grooves in the spine into the wound.

Watch your step

Although stingrays can live on the sea floor, hundreds of metres below the surface, they can also be found in shallow waters near beaches and in estuaries. This makes them quite dangerous to humans. It's very easy to accidentally stand on a stingray buried or half-buried in the sand in shallow water. If this happens, the stingray's tail shoots upwards and it will thrust its barbed spine (see Figure 12-5) into you, leaving a deep and painful wound.

The venom injected by a stingray barb can cause nausea, vomiting and muscle cramps, but it's not normally life-threatening. The wound is likely to bleed profusely and fragments of the barbed spine often break off in the wound, leading to later infection. The greatest dangers of a stingray sting are blood loss from the wound or if the barb punctures a major organ (see the following section to find out why). Fortunately, most stings strike people's feet, lower legs, hands or arms.

Soak the wound in very warm water to relieve pain and help neutralise the venom. Try to remove any fragments of the barbed spine from the wound, but only if this is unlikely to cause further damage or bleeding — don't attempt to remove any fragments embedded in the chest or abdomen. Also, don't use a pressure immobilisation bandage or try to close the wound, but apply pressure to bleeding wounds in the case of arterial damage. Seek medical treatment as quickly as possible. You may also require a tetanus injection.

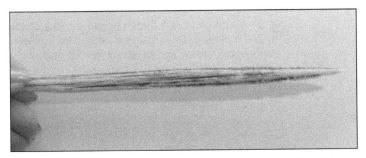

Figure 12-5: The barbed spine of a stingray can cause deep and painful wounds.

If the barb has been removed and bleeding is profuse, you can apply some pressure with a towel while waiting for medical assistance. If the wound is to a limb, raising it can help reduce blood loss.

The deadliest attacks

Despite their non-aggressive nature, stingrays have been responsible for three recorded human deaths in Australian waters. In each of the three attacks, a stingray's barbed spine pierced the heart of the victim. The most recent fatality occurred in September 2006, when 'Crocodile Hunter' Steve Irwin was tragically killed while filming a documentary in the tropical waters off Port Douglas in northern Queensland. Irwin was swimming above a stingray when it suddenly thrust its tail upwards, driving its barbed spine into his heart. The attack was unprovoked. It's most likely that the stingray either mistook Irwin for a predator, like a shark, or simply felt cornered and was trying to defend itself.

Rays with a z-z-zap

Two rays found in Australian waters deliver electric shocks from organs in their disc-shaped bodies, rather than use a barbed spine on their tails, when defending themselves from predators. Like other rays and stingrays, the Short-tailed Torpedo Ray and the Numbfish have skeletons of cartilage and glide smoothly through the water. Both of these rays bury themselves in sand on the sea floor and use a jolt of electricity to stun their prey as it passes by.

  ✔ The Numbfish can live in ocean depths of more than 200 metres (219 yards), as well as in shallower waters near the shore and in estuaries. To find out more about the Numbfish, refer to Chapter 10.

  ✔ The Short-tailed Torpedo Ray, also known as the Electric Ray or Torpedo Ray, is found in deep water all around the Australian coast (including Tasmania), except for the tropical north. This ray grows to more than 1 metre (3.3 feet) in length, and is grey, brown or yellow on top and white underneath. The Short-tailed Torpedo Ray has been known to deliver powerful electric shocks to divers and anglers. Although the electric shock is unlikely to directly cause death, if the victim loses consciousness in the water, drowning could possibly occur.

The electricity generated by these rays is produced by their muscle cells, which work together like the electric cells in a car battery. They cause the top surface of the ray to become positively charged, while the bottom surface becomes negatively charged. When contact with another organism is made in saltwater, an electric current passes through it, delivering an unexpected electric shock.

Remove the victim from the water straight away. Keep the victim calm and warm, and seek medical attention. In severe cases, the victim may stop breathing. Apply mouth-to-mouth resuscitation and external heart massage if necessary (see Appendix A) and call an ambulance.

Smaller Fry in Deep Water

Some dangerous deep-water creatures are much smaller than sharks and stingrays. When fishing in deep water, treat any fish with one or more spines with caution. Even the common flathead, popular with recreational anglers all around the Australian coast, should be handled carefully. Its venomous spines can cause painful stings, which though not directly life-threatening, can lead to painful infections.

The most dangerous fish with spikes are the scorpionfish, and in the following sections we look at two deep-sea dwellers — the Bighead Gurnard Perch and the Fortesque. (Refer to Chapter 11 for information on scorpionfish that live in shallower waters.)

It's a killer: The Bighead Gurnard Perch

One of the most dangerous of the scorpionfish lives in water at depths of between 15 and 1,200 metres (16 to 1,300 yards), so you're unlikely to encounter it while swimming at the beach. However, you could accidentally encounter the Bighead Gurnard Perch while diving near reefs or angling. This scorpionfish (shown in Figure 12-6) is less than 50 centimetres long (20 inches) and is found off the coast of southern Western Australia, South Australia and Tasmania.

You have to admit it — this fish looks pretty scary. Those venomous spines can cause painful puncture wounds that can easily become infected. Even when dead, this fish can cause serious wounds. The venom injected can cause nausea, vomiting, muscle weakness and breathing difficulties. The Bighead Gurnard Perch has contributed to two deaths in Australian waters.

Soak the wound in very warm water to help neutralise the venom and relieve pain. Seek medical treatment as quickly as possible.

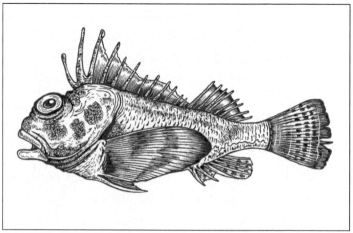

Figure 12-6: The Bighead Gurnard Perch can inflict lethal wounds.

Spiky bottom dwellers: The Eastern Fortesque

Along the east coast, from southern Queensland to Victoria, anglers are occasionally injured by a bottom-dwelling scorpionfish called the Eastern Fortesque (see Figure 12-7). This scorpionfish has some close relatives that look and behave similarly, and are equally dangerous. These include the Marbled Fortesque, which lives in Queensland and New South Wales, and the Western Fortesque, found off the south coast of Western Australia.

Like other members of its genus, the Eastern Fortesque only grows up to 15 centimetres (6 inches) long and lives in estuaries and bays at depths of up to 30 metres (33 yards). However, these little fish have 16 venomous spines on their backs and more on their fins. These spines can cause painful stings.

Fortesques lurk on the sea floor, sometimes in large numbers, laying in wait for passing prey. Their colour and markings camouflage them well, so they're experts at surprise attacks.

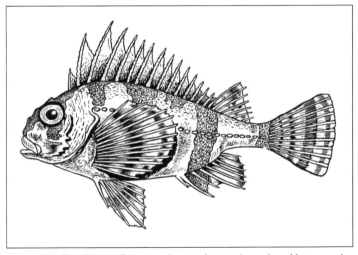

Figure 12-7: The Eastern Fortesque is one of several species of fortesque in Australian waters.

An Eastern Fortesque's sting is very painful but its venom is not a serious threat to humans. The most frequent victims of Fortesque stings are amateur divers, who, unaware of the danger, see this slow-moving fish as an easy catch, and anglers who fail to wear sturdy gloves while trying to remove the fish from their lines or when sorting their catch in fishing nets.

Relieve pain by immersing the wound in very warm water. If any other symptoms are present, seek medical attention. Clean and bandage the injury. Check the wound daily for signs of infection (see Appendix A).

Part IV
Urban Living

Glenn Lumsden

'The next no-good bum that messes up my house is going to get such a bite!'

In this part . . .

A home in the wild? No, this part looks at wildlife around the home . . . and in the garden. You may think you're safe in your own home, but you'll probably think again after you read these chapters.

There's myriads of dangerous creatures that like to take shelter in or around dwellings. Australian homes and gardens provide ideal hiding places for many animals. For example, snakes, Cane Toads, birds and bees just love gardens. Rats and spiders like the shelter of homes, and bedbugs and fleas just love people's bodies.

Everything's okay, though. In this part we give you tips and advice on how to prevent dangerous encounters with these animals.

Chapter 13

Along Came a Spider

*T*he most likely place to find spiders is in your own home or garden. Many people are terrified of all spiders, whether large or small. But you're probably more likely to injure yourself trying to avoid a harmless spider than by being bitten by a dangerous one. Imagine, for example, a large and harmless Huntsman Spider dropping into your lap when you pull the car's sun visor down — this does happen, and it's caused some serious road accidents.

Luckily, the bite of most Australian spiders produces no more than local swelling at the site of the bite, some pain and itchiness later on. However, certain species produce extremely toxic venom that can cause severe illness or even death. In this chapter, we show you how to identify Australia's deadliest spiders, tell you how to avoid deadly encounters with them and cover how to deal with spider bites — if you're bitten. We also give you some good news; spiders are fascinating creatures that you can admire from a distance. If you leave them alone they won't harm you.

 Australia is home to more harmless spiders than dangerous ones, but we recommend that you treat all spiders with caution. Wear gloves and shoes when you're gardening, handling piles of rubbish and shifting building materials.

If you're bitten, try to identify the spider, either by observing it directly or by capturing it (as long as this doesn't endanger you further). If you or someone else catches the spider, place it in a jar with some methylated spirits. If you think that you've been bitten by one of the harmful spiders mentioned in this chapter, take it along with you to the doctor or hospital. A positive identification will allow the correct treatment to be administered more quickly.

Seeing Red

The Red-back Spider bites more people each year than any other Australian venomous creature (about 300 bites per year). Before 1956, the Red-back was regarded as Australia's most dangerous spider, killing at least 14 people. Since then, an antivenom has been available and only one confirmed death from a Red-back Spider bite has been recorded.

Females are the deadly members of the species (see Figure 13-1 and the photo in colour section). They vary in body length from about 6 to 15 millimetres (0.2–0.6 inches) and have a distinctive red pattern on their abdomen. Their leg span is up to 30 millimetres (1.2 inches).

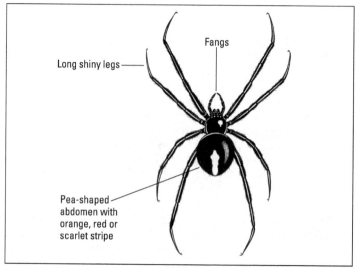

Figure 13-1: A female adult Red-back Spider.

Slim Newton's ode to the Redback

The 1972 hit song, *Redback on the Toilet Seat*, by Slim Newton, could almost have been based on fact. Red-back Spiders like the dark corners and shelter that would have been provided by the now rare (or even extinct) Australian backyard toilet. The only problem is that the song blames a male spider for the bite. But it could only have been a female.

Male Red-back Spiders are much smaller than females (about 3 millimetres or 0.1 inches in body length) and have lighter coloured markings. Although male Red-backs are venomous, their bite is unlikely to break human skin.

Look before you touch

Red-back Spiders are found in urban environments all over Australia, but rarely in the bush. Female Red-backs construct their untidy webs in dry and sheltered areas such as wood-piles, garden sheds and the undersides of tables and chairs. Even a pair of shoes or gumboots left outside can provide an ideal location for a Red-back web. Remember to look before you touch or sit on anything that may contain a Red-back Spider's web.

About the bite

Red-back Spiders are not aggressive, and usually roll into a ball and fall to the ground, feigning death, if threatened. Most bites occur accidentally, when the spider is picked up or trapped against a hand or body.

Although the female Red-back only injects a small amount of venom, a bite can cause serious illness, resulting in pain, inflammation and swelling at the site of the small puncture wound. Sometimes, more severe symptoms can occur: Nausea, vomiting, increased blood pressure, patchy sweating, aching joints and extended pain. In the most severe cases the venom can cause paralysis, which may lead to death. Children are more at risk from the effects of the venom because of their smaller size.

Making a meal of males

The Red-back Spider (*Latrodectus hasselti*) is closely related to the notorious Black Widow Spider (*Latrodectus mactans*) of North America. The Black Widow gets its name because the female makes itself a widow by eating the male after mating. Like its relative, the female Red-back eats the male after mating.

Mating is also a hazardous event for a number of harmless Australian male spiders — for example, orb spiders, fishing spiders, wasp spiders, flower spiders and garden spiders. It may seem brutal for a female to eat its mate, but this important food source provides protein and energy for the female to lay her eggs, build her egg sac and care for her *spiderlings* (offspring).

Eating the mating male is not unique to the spider world; praying mantises, scorpions and some species of biting midges and crickets are also known to eat their male mating partner.

The recommended first aid for a Red-back Spider bite is to apply a cold pack to the site of the puncture wound. Don't apply a pressure bandage, because this is likely to increase pain around the wound. Keep the victim calm and take him or her to a doctor or to hospital, along with the spider — if possible.

The Infamous Sydney Funnel-web

The Sydney Funnel-web is one of Australia's largest and most easily identifiable spiders (see photo in colour section and Figure 13-2). It's also the most dangerous of about 40 known species of funnel-web spider, which all live in different places along the southeast coastline of Australia — from the Daintree Rainforest in north Queensland to the open forests of the Eyre Peninsula in South Australia. (For information about these funnel-web spiders, see the section 'All in the family: Other dangerous funnel-web spiders' later in this chapter.) If you ever see a funnel-web spider, you'll recognise it as being dangerous, instantly: It's big, it's black, it's hairy and really ugly. Its fangs are also very long and sharp, so you'll know if you're bitten by one, too.

Funnel-web spiders have a pair of prominent *spinnerets* extending from the end of the abdomen. Spinnerets are usually small appendages at the base of a spider's abdomen through which the silk is extracted to make the spider's web, but on funnel-web spiders the spinnerets are larger to allow them to spin their funnel-shaped webs.

When threatened, funnel-web spiders rear up on their hind legs, displaying their fangs, dripping with venom. The male is slightly smaller than the female. But unlike the Red-back Spider, it's the male that's most dangerous. The male funnel-web spider's venom is about five times more toxic than the female's.

Sydney Funnel-web Spiders can be found anywhere within a radius of about 120 kilometres (75 miles) of Sydney — usually in suburban gardens.

They build their tube-like nests of thick silk in the crevices of rocks, in stone walls, logs or even in the ground — so it's easy for humans to accidentally expose themselves to these spiders while gardening. Oddly enough, dogs and cats are hardly affected by funnel-web venom.

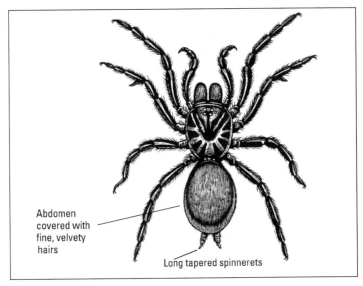

Abdomen covered with fine, velvety hairs

Long tapered spinnerets

Figure 13-2: A male adult Sydney Funnel-web Spider.

The wanderers

Female Sydney Funnel-web Spiders rarely leave the web, but males roam around in summer and autumn looking for a mate. They sometimes mistakenly venture inside a house, where they may hide in clothing and bedding. If disturbed, they rear up and deliver a bite in self-defence. Fortunately, these male adventurers usually die of thirst before causing any trouble.

Male funnel-web spiders can also wander into swimming pools and sandpits. If you live in the Sydney area, check these areas around the home before allowing children to use them.

According to the Australian Museum, funnel-web spiders can float on a pool for up to 44 hours, and can survive between 24 and 30 hours under water, but by then they're incapable of biting for an hour or so.

A deadly bite

In most bite cases, the Sydney Funnel-web Spider isn't able to inject enough venom into its victim to cause death. However, its venom is extremely toxic. At least 14 people have died as a result of a bite from a Sydney Funnel-web Spider. The venom works very quickly, producing violent nervous reactions within 15 minutes of the bite. Several small children have died within two hours of being bitten. In adults, death usually occurs within three days.

The earliest symptoms of a Sydney Funnel-web Spider bite include:

- ✔ Localised pain around the region of the bite
- ✔ Tingling and numbness, especially in the mouth
- ✔ Sweating
- ✔ Abdominal pain
- ✔ Nausea and vomiting
- ✔ Salivation

If the bite is on a limb, apply a pressure immobilisation bandage (see Appendix A to find out how). Restrict movement to the limb or the area where the bite is located, to slow down the spread of the venom. Keep the victim calm and call an ambulance (provide the spider too — if caught).

All in the family: Other dangerous funnel-web spiders

All funnel-web spiders look similar to one another and behave in similar ways to their Sydney relative. Some of them are dangerous because their venom has similar effects to that of the Sydney Funnel-web. However, only the Sydney Funnel-web has caused fatalities in Australia. Other dangerous funnel-web spiders include:

- ✔ The Northern Tree Dwelling Funnel-web (southern Queensland and northern New South Wales)
- ✔ The Southern Tree Dwelling or Paper-bark Funnel-web (south of the Hunter River, New South Wales, and common around Sydney)
- ✔ The Blue Mountains Funnel-web (all of New South Wales)
- ✔ The Darling Downs or Toowoomba Funnel-web (southeast Queensland and all of New South Wales)

These spiders can be found in gardens or indoors in areas in or near their natural habitat.

Distinguishing between the different species of funnel-web spider is really difficult, so always consider a bite from any spider suspected of being a funnel-web as dangerous. Seek medical help straight away if you notice any signs of tingling or numbness, sweating, abdominal pain, nausea, vomiting or salivation (refer to the preceding section for more details). And keep the spider to assist with identification, if possible.

How to catch a spider

How do you catch a spider? Carefully. Remember, most spiders have venom, even though they rarely bite.

Chasing and trying to trap a spider will upset it, so you must take care: The spider will try to defend itself. If you're in danger of being bitten, don't even try to capture it. Instead, take a photo (preferably digital, for convenience) or just attempt to get a good look at it so that you can provide a good verbal description.

If you're confident you can catch the spider without endangering yourself or anyone else, follow these guidelines:

You need

✔ A small clear glass jar and lid

✔ A piece of cardboard or stiff paper — a bit bigger than the top of the jar

✔ A pair of gardening gloves

Now, follow these steps:

1. Put on your gloves, then isolate the spider on a flat surface — the floor, wall or table top.

2. Place the jar over the spider and hold it flat with the surface (so that the spider has no escape route).

3. Slip a piece of cardboard or stiff paper between the jar and the surface the spider is sitting on by lifting the jar only slightly.

4. Turn the jar upright with the cardboard still on top and tap the jar until the spider drops.

5. Secure the lid on the jar.

Well done! You've captured the spider.

Spiders with Attitude

Most spiders are not aggressive towards humans. But some *become* very aggressive and frightening if you dare to disturb them or their nest. The spiders described in the following sections can give you a very nasty bite.

Trapdoor spiders

The name 'trapdoor spider' is used to describe a variety of spiders that nest in web-lined burrows, often covered by a trap door made of silk and soil. The camouflaged door is hinged so that when prey is within range of the burrow, the spider can rush out, grab its prey and drag it back inside its burrow to eat, dropping the door back in place without delay. Trapdoor spiders vary in body length from about 10 millimetres (0.4 inches) long to about 50 millimetres (2 inches) long and are often mistaken for funnel-web spiders. However, the spinnerets of the trapdoor spiders are shorter and less pointy than those of the funnel-web (see Figure 13-3).

Do not disturb

Although trapdoor spiders aren't normally aggressive, they have large fangs and can deliver a painful bite when provoked. With the exception of the Mouse Spider (see the next section), none of the trapdoor spiders is considered to be dangerous. Pain and swelling are usually the only effects of the venom. But in a very small number of cases, victims have become quite ill.

 If you're bitten by a trapdoor spider, try to catch it (taking care not to be bitten again!), just in case you do become ill and identification of the spider is required.

 You can apply a cold pack to the site of a trapdoor spider puncture wound to relieve pain and swelling. No further first aid treatment is required unless illness develops. Keep the spider in case you need to seek medical treatment.

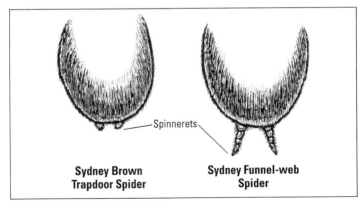

Spinnerets

Sydney Brown Trapdoor Spider

Sydney Funnel-web Spider

Figure 13-3: The spinnerets of trapdoor spiders, which they use to spin a web to line a burrow, are shorter and less pointy than those of funnel-web spiders.

They're everywhere!

Trapdoor spiders can be found throughout Australia. Their common names often reflect the areas that they inhabit.

- Melbourne Trapdoor Spider
- Sydney Brown Trapdoor Spider
- Tasmanian Golden Trapdoor Spider
- Nullarbor Cave Trapdoor Spider

The trapdoor name is a little misleading because not all trapdoor spiders make burrows with lids. That is, there's no trapdoor — the spiders simply live in web-lined burrows in the ground. Even more confusing, some experts describe the Mouse Spider as a trapdoor spider, while other experts don't.

Just call me 'Fang': The Mouse Spider

The Mouse Spider (see Figure 13-4) is often described as a trapdoor spider because it builds burrows with trapdoors in the ground. Its deep burrows are often found in urban lawns and gardens. When disturbed, Mouse Spiders become angry and aggressive. They rear back as they prepare to attack with their venom-filled and menacing fangs. Mouse Spiders are found throughout mainland Australia.

Mouse Spiders can vary in colour but, overall, they're recognisable because they have a shape similar to the funnel-web spider. This is probably a good thing, because their venom is toxic and produces similar symptoms. Although no fatalities have been recorded, Mouse Spider bites should be treated as seriously as those from a funnel-web spider. The most serious bite case on record was successfully treated with Sydney Funnel-web antivenom.

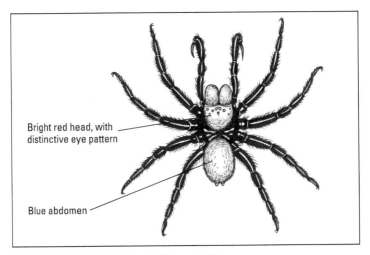

Bright red head, with
distinctive eye pattern

Blue abdomen

Figure 13-4: A male Red-headed Mouse Spider.

If a bite from a Mouse Spider is sustained on a limb, as
it usually is, apply a pressure immobilisation bandage
(see Appendix A for details). Restrict movement to the area
of the bite to slow down the circulation of venom. Keep the
victim calm and take to a doctor or hospital, along with
the spider, if possible. Antivenom will be administered, if
necessary.

The huge and hairy Australian tarantula

It's huge! No — make that ginormous! A spider the size of
a dinner plate! The Australian tarantula, also known as the
barking spider, bird-eating spider or whispering spider,
looks very scary — see Figure 13-5. Its body is typically
about 60 millimetres (2.4 inches) long, with a leg span of
about 160 millimetres (6.3 inches). The 'bird-eating' name is
deceptive, though. Australian tarantulas mostly eat insects
and other spiders — and perhaps the occasional small frog
or lizard.

Figure 13-5: An Australian tarantula ready to attack.

Despite their scary appearance, Australian tarantulas aren't normally aggressive. But, like funnel-webs and Mouse Spiders (covered earlier in this chapter), they rear up and bite when disturbed or threatened. That's when they make a barking or whistling sound. They live in burrows lined with their web and can be found throughout most of mainland Australia, although some species of the tarantula are only found in Queensland and northern New South Wales.

Because the Australian tarantula is such a big spider, with big fangs, its bite is extremely painful. The spider's venom can cause nausea and vomiting. Until recently, Australian tarantula bites were rare, but because more and more people are now keeping them as pets, bites are more common.

Apply a cold pack to the site of the tarantula bite puncture wound to relieve pain and swelling. No further first aid treatment is required unless illness develops. (Keep the spider in case it's necessary to seek medical treatment.)

Don't be afraid of the big bad Wolf Spider

Wolf Spiders get their name because they chase and run down their prey — much like wolves do. But you shouldn't be afraid of them: Humans are a little large for them to consider as dinner.

However, Wolf Spiders (see Figure 13-6) can be aggressive towards humans if they're approached. And they will bite if you make them angry. A bite from a Wolf Spider has been responsible for the death of an adult Labrador dog, whereas in humans, medical experts suspect (but have not yet confirmed) that a Wolf Spider bite can produce skin ulcerations and possibly cause gangrenous infections and severe kidney breakdown.

If a Wolf Spider strikes, apply a cold pack to the site of the puncture wound to relieve pain and swelling. No further first aid treatment is required unless illness develops. Keep the spider in case you need to seek medical treatment.

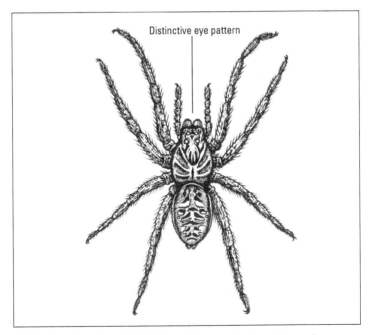

Distinctive eye pattern

Figure 13-6: The Wolf Spider's distinctive eyes allow it to observe its surroundings and prey with great accuracy.

Creepy, But Not So Nasty

Almost all spiders are venomous, but very few can really be described as dangerous to humans. Of the Australian spiders, only the Red-Back Spider and the Sydney Funnel-web Spider (covered earlier in this chapter) are known to have caused deaths since records began being kept in 1927. Mouse Spiders and Australian tarantulas can also cause serious illness.

The venom of most spiders is only toxic enough to kill insects and other small creatures that they eat — including other spiders. But the venom of several other spiders that you may find at home *could* be described as dangerous — their bites can sometimes cause illness and skin conditions. It's also possible that puncture wounds caused by relatively harmless spiders can become infected.

Spiders in general aren't to be fooled with. If you spot one, avoid contact and just watch the spider — and admire the beautiful web — from a safe distance. Precaution is better than cure; but, if you're bitten, try to identify the spider and keep a close eye on the bite site for any changes to the area, and any other signs of illness. When in doubt — especially if the victim is a child — get the bite checked by a doctor.

To observe spiders, view them from a safe distance and, if possible, behind glass. You'll feel a sense of wonder and respect for these tiny creatures when you:

- ✔ Look at the overall body plan. Notice its shape, size, patterns, colours, length and thickness of legs, hair or bristles, and the number and layout of the eyes.

- ✔ Look at a web and notice its ingenious construction.

- ✔ Try to observe your spider at night — that's when it's most active; repairing its web and capturing prey.

- ✔ Take enough time to enjoy the engineering in the spider's work and its cunning when capturing food.

- ✔ Look for evidence of the spider's last meal, egg sacs and spiderlings hatching.

The White-tailed Spider — an undeserved reputation

The White-tailed Spider, shown in Figure 13-7, has a reputation for being quite dangerous. Since the early 1980s, its bite has been associated with severe skin necrosis and ulceration. However, many experts now believe that there's not enough evidence to conclude that these symptoms are actually caused by the bite of a White-tailed Spider.

Necrosis is a medical term used to describe the death of cells in the tissues. In the case of spider bites, it can be a reaction to the spider's venom on skin cells or due to the bite getting infected with a disease-causing bacteria. Burns and the blocking of arteries supplying a tissue also causes necrosis.

Figure 13-7: White-tailed Spiders are quite easy to identify.

Tickling for treats

White-tailed Spiders have an unusual way of catching some of their prey. They tickle the Black House Spider's web until the hapless victim comes out to investigate the disturbance. As soon as the Black House Spider emerges, the White-tailed Spider pounces.

White-tailed Spiders live in all regions of Australia. They don't spin webs, but wander in search of shelter and prey. Bathrooms, laundries and bedrooms provide White-tailed Spiders with both shelter and a supply of their preferred food — other spiders. They like the warmth and protection of clothing and bedding; so you're most likely to be bitten when putting on clothes or when slipping into bed.

A bite from the White-tailed Spider can be quite painful, but in most cases the only other symptoms are local swelling, redness and itching.

A cold pack applied to the site of the bite can relieve pain and swelling. No further first aid treatment is required unless illness develops. Keep the spider in case you need to seek medical treatment.

House guests or pests?

Some of the relatively harmless spiders that you're likely to encounter in homes and gardens can bite when disturbed, though many people don't worry about them. The following spiders like to dwell in or near homes, but have a 'creep' factor for one reason or another:

- ✔ **Black House Spider (Window Spider):** This common little 1–1.5 millimetre (0.04–0.06 inch) black spider is often called the Window Spider because it usually builds its untidy webs in the corners of windows or the crevices between bricks. People bitten by the Black House Spider have complained of severe pain and swelling around the bite, sometimes leading to sweating and vomiting.

Daddy Long-legs: A legend in his own lunchtime

It's widely believed that the Daddy Long-legs Spider (pictured) is the most venomous spider of all, but that its fangs are too short to penetrate human skin. This myth is wrong on both counts. Firstly, there's no scientific evidence that the Daddy Long-legs' venom is highly toxic. Even if it were, the amount of venom released in a bite would not be enough to cause harm to a human. Secondly, its fangs can be long enough to penetrate human skin on some areas of the body, but its jaws are unable to force them into it. The myth may have come about because Daddy Long-legs Spiders do kill and eat the much more dangerous Red-back Spider.

✔ **Australian Fiddle-back Spider:** This spider is shy and retiring, and found only in the suburbs of Adelaide. The Fiddle-back Spider's defining characteristic is its light brown body with the outline of a fiddle or violin shape on its upper body. It has long legs, is very reclusive, and likes to hide out in dark, moist places around the house. Measuring 15 to 20 millimetres (0.6 to 0.8 inches) when adult, this spider is thought to be the culprit of a number of severe illnesses attributed to spider bites.

✔ **Huntsman Spider:** The Huntsman looks ferocious, but rarely bites when disturbed. In fact, the Huntsman usually runs away. The Huntsman Spider ranges in colour from dark brown to light brown and can grow as large as 15 centimetres (6 inches). It has long legs that are sometimes banded light and dark brown. This spider moults in order to grow, often leaving its spent-off skeletons as evidence of its presence. Some rare cases of bites causing illness from a Huntsman have been recorded.

Chapter 14

Out and About

*Y*ou may consider local parks and reserves, or your own garden, great places to spend time in. And you're right, especially if you want to kick a ball or sit down and read a book. However, if you poke your nose, hands or feet into the wrong places, or wander into the path of an animal keen on defending its young, you're likely to come off second best.

In this chapter, we look at the creatures you're likely to meet when you venture outdoors in urban areas.

Suburban Stingers

Deaths caused by sharks and crocodiles always hit the headlines. But common garden creatures, such as bees, ants and wasps, which live around homes and in other urban areas, are the most deadly. Since 1979, more than 70 people have died in Australia after receiving stings from these insects.

Busy bees

Most Australian native bees are solitary little fliers and, being vegetarian, they normally wouldn't hurt a fly. They build their nests in out-of-the-way places — in dead trees or underground — and prefer to buzz around solo, collecting pollen and nectar undisturbed by other creatures, including their own species. Some Australian native bees have no stings; but others do, and boy, can they sting if you get in their way!

Stay away from the hive, Honey!

The industrious little Honey Bee is the species of bee that's most likely to 'wander' into urban gardens looking for nectar. These bees are more sociable than other types of bee. They live in large groups in nests called *hives*, and when threatened release a pheromone that attracts other Honey Bees. One lone, threatened bee in a garden can deliver a nasty or even fatal sting, but if you disturb a hive the other bees quickly respond to help out in the attack.

A *pheromone* is a chemical message or scent, made by an organism to attract or warn members of the same species.

To avoid being stung by a bee, always wear shoes when you're walking on grass. Also, to ensure you're not attractive to a bee, don't wear brightly coloured clothes or a floral-scented perfume.

Suicide attack

Female Honey Bees (called worker bees) have a barbed sting with an attached venom gland at the tail end of their body.

(See Figure 14-1 and photo in colour section.) In the bee's frenzy to escape from you, it twists and turns, leaving the barbed sting and venom gland behind, fatally wounding itself. The bee soon dies without its sting and venom gland, and the wound you receive throbs and throbs as the venom pours from the gland into your system.

Male Honey Bees (called drones) don't have a sting. Their only role is to fertilise the queen bee. They don't go out in search of nectar — the females worker bees do all the collecting. Bee hives usually have only one queen bee.

Avoid drinking directly from a soft drink can when you're outside — use a straw. Honey Bees, attracted to the sugar, have a habit of flying into open soft-drink cans. If you're not looking when they do, your next sip may include a bee — and the bee won't be impressed! A sting to the back of your throat may cause swelling and choke you.

Allergies

Many people are allergic to bee stings. The allergic reactions vary from minor to life-threatening.

Barbed sting

Figure 14-1: The Honey Bee and a close-up of its barbed sting.

Allergic reactions can include:

✔ A rash or redness developing around the site of the sting

✔ Swelling of the tongue or throat

✔ Difficulty in breathing

✔ Stomach cramps

✔ Vomiting

✔ Diarrhoea

If you've never been stung by a bee, or you've been stung only once before, you may not know whether you're allergic to the Honey Bee's venom.

Remove the bee's sting immediately, by either scraping it off the skin or quickly pulling it out. After the sting is removed you can use a cold pack to relieve pain. If any signs of an allergic reaction are present, seek medical help urgently. (See Appendix A for information on how to proceed if symptoms become life-threatening — for example, if the patient has difficulty breathing or collapses.)

Ant antics

Ants live just about anywhere in Australia. Luckily, most of them aren't at all dangerous. The ants that you're likely to find around homes or buildings are generally tiny and too busy building nests and collecting food to be bothered with humans. They can be a nuisance when they invade your house or garden in large numbers, though. However, in areas closer to bushland, you're likely to be confronted by larger and more dangerous ants.

Bull ants: Jaws on six legs

Bull ants (see Figure 14-2) are ants with attitude. Markings between bull ant species vary, but all have powerful jaws and a nasty sting in their tail. If you get too close to a nest, they come charging out looking for trouble. They'll grab onto you with their jaws, coil their body into a spring and deliver a powerful and painful, venom-loaded sting. If you don't flick the ant off quickly, it can sting you again and again — releasing a pheromone to invite its mates along for the feast. If you're allergic to their venom, a single sting can be fatal.

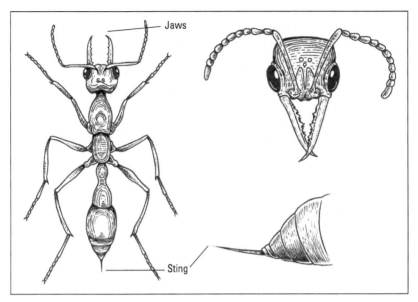

Figure 14-2: The bull ant has big jaws and a nasty sting in its tail.

Bull ants are among the largest ants in the world, measuring up to 30 millimetres (1.2 inches) in length. They live in large nests several metres below ground, with small entrances at surface level. Many different species of bull ant can be found throughout Australia. Some of them are brightly coloured, with yellow, red or orange markings on their heads or abdomens — see the photo in colour section. Others are very plain to look at — brown or black. But whatever their colour, there's no mistaking their jaws.

Jumping Jacks

Jumping Jacks (also known as Jumper Ants) are smaller than bull ants, but more aggressive. They range in length from about 10 to 14 millimetres (0.4 to 0.6 inches) and have bright yellow or orange jaws and legs. Jumping Jacks aggressively jump towards animals, including humans, that get in their way. Like bull ants, they bite and coil themselves up into a spring to deliver a sting with a powerful punch. Jumping Jacks (see Figure 14-3) are found throughout southern Australia — in regions that include most of the country's major cities and towns.

Figure 14-3: A Jumping Jack carrying its lunch.

Treating ant stings

Stings inflicted by bull ants and Jumping Jacks cause pain or swelling around the area of the wound. In most cases, no further symptoms occur, although you may experience discomfort for a day or two. However, if you're unlucky enough to be allergic to the venom, more serious symptoms can develop, including difficulty in breathing, swelling in other areas of the body away from the wound, dizziness and loss of consciousness. If not treated, allergic reactions can cause death.

Apply a cold pack to the site of the sting to reduce local pain and swelling. If you note any signs of an allergic reaction, such as difficulty in breathing, vomiting, dizziness or unconsciousness, seek medical help urgently. See Appendix A for information on how to respond if symptoms become life-threatening.

Native wasps to watch out for

Australian native wasps, like their European cousins (see Chapter 15), are more than capable of spoiling a sunny afternoon outdoors. Suburban parks, homes and gardens provide an ideal shelter for their nests. In this section, we show you three native varieties to avoid.

Wasp stings aren't barbed so, unlike bees, they can sting more than once — and they certainly will do so if you provoke them.

The Blue Ant: A wasp without wings

The Blue Ant is a wingless wasp, not that this technicality is your main concern. If it bites you, it stings like billyo! The Blue Ant has a metallic blue body some 30 millimetres (1.2 inches) long and resembles a bull ant, but with a waist — see Figure 14-4. Like bull ants, it delivers a fearsome bite or multiple bites, given the opportunity. Blue Ants nest in suburban parks and gardens, and if you accidentally stand on or dig up their nest, they become as aggressive as an angry bull.

Paper and spider wasps

Several species of paper wasp and spider wasp are commonly found in Australian suburban gardens. These wasps have four wings and are more slender than bees. They can attack and sting you, but only if you threaten them.

Figure 14-4: The Blue Ant — a wasp without wings.

✔ **Paper wasps:** These wasps (see Figure 14-5) are black, with yellow or orange bands on their abdomen. Paper wasps are social animals — they nest in large colonies. They construct nests that resemble grey papery material, usually under eaves and verandas or high up in trees. Their nests are delicate and very beautiful, made up of perfect hexagonal cells (see photo in the colour section). The nest starts off cone-shaped, and becomes rounder and rounder as the colony grows and more cells are added. Looking at these nests is okay, but don't get too close, otherwise the wasps may invade your space, en masse — and painfully. They are also potentially lethal.

Don't ever attempt to remove a paper wasp nest yourself. Instead, call a pest control company or the local council.

✔ **Spider wasps:** Common in suburban backyards all over Australia during summer, spider wasps (see Figure 14-5) are large, black and orange. They're solitary wasps that nest in mud burrows and feed on spiders — including those larger than the wasp, such as the Huntsman Spider. Ranging in size from a few millimetres to around 35 millimetres (1.4 inches), spider wasps look dangerous, but they're usually not aggressive. Leave the spider wasp alone and it will leave you alone.

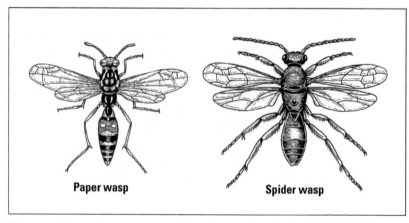

Paper wasp   Spider wasp

Figure 14-5: Australian native wasps.

Dealing with wasp stings

Wasps can deliver nasty stings that often swell up and look like welts. To relieve pain and swelling, apply a cold pack to the site of the sting. More serious symptoms, such as breathing difficulties, stomach cramps, vomiting, dizziness or unconsciousness, indicate an allergic reaction and require immediate medical attention. (See Appendix A to treat life-threatening symptoms.)

Assassins in the garden: Assassin bugs

The name assassin bug sounds ominous, doesn't it? Assassin bugs, no matter which species, are killers of other insects and sometimes cannibalise their own. (See Figure 14-6.) They're slow moving and stealthily ambush their quarry — grabbing and holding prey with their legs, then piercing into flesh with their biting parts before sucking out the body fluids.

A _proboscis_ is a long, tube-like mouthpart used by insects for piercing or sucking food.

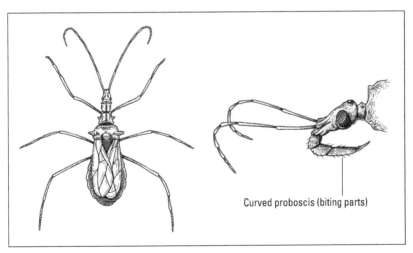

Curved proboscis (biting parts)

Figure 14-6: Assassin bugs have a distinct neck between the head and thorax, and a proboscis that curves backwards.

Although assassin bugs don't usually attack humans, you need to take care when you encounter them because these insects can deliver a painful bite. They inject a salivary fluid into the victim, but it's not toxic.

If you're bitten by an assassin bug, apply a cold pack to the site of the sting to reduce local pain and swelling.

Jeepers Creepers

Some caterpillars look hairier than others, but the hairy varieties are the type that you need to be most wary of. If you touch, or brush against one of these caterpillars, you're likely to be stung by the pointy hairs.

Some hairy caterpillars, like the one shown in Figure 14-7, are venomous. The most common venomous caterpillars are the Gumleaf Skeletoniser and the chinese junk varieties. These caterpillars grow into harmless adult moths — the Gumleaf Skeleton Moth and cup moths respectively. As caterpillars, they live on the leaves of trees — especially eucalypts — almost anywhere in Australia, apart from the Northern Territory.

Figure 14-7: The brittle hairs on hairy caterpillars are left embedded in your skin if you touch them.

The hairs on hairy caterpillars resemble the spikes you might find on a cactus plant — pointy and brittle. Glands at the base of these hairs produce venom that can cause severe skin irritation.

If your skin makes contact with a hairy caterpillar, it leaves venom-releasing hairs embedded in your skin. Most hairs can be washed off. You can remove stubborn hairs by attaching transparent sticky tape to the affected area and stripping them off.

Usually, a cold pack applied to the site of stings can reduce local pain and swelling. Keep an eye on the victim for symptoms of severe allergic reaction (see Appendix A). If embedded hairs can't be washed or stripped off, or if irritation continues for a long period, seek medical advice.

Dangerous Dogs

About four million dogs are kept as pets in Australia. The statistics available on dog attacks are incomplete and fragmented, since no comprehensive reporting system exists. However, each year more than 100,000 Australians suffer dog attacks, causing injuries of varying degrees of severity. Almost 1,400 of these attacks cause injuries serious enough to warrant hospitalisation. Fortunately, the mortality rate is low, but dog bites can still cause severe physical and emotional damage.

Around 60 per cent of all serious bites occur to children under the age of 10, and many of these bites are to the head, face and neck. Adults are more likely to be bitten on the lower limbs and hands.

Most dog attacks occur in the home environment, or in the home or backyard of a friend, neighbour or family member.

Although most dogs don't bite, they can be unpredictable, so it's sensible to avoid risks with dogs you don't know, and to remain wary of dogs you do know. Responsible dog owners register their dogs and ensure they don't stray from their yards unless supervised and on a leash.

Here are some ways to reduce the risk of an attack:

- Never approach a strange dog unless the owner is at hand, and the dog has been properly introduced and is friendly.
- Stay well out of reach of a dog on a chain or in a car.
- Never tease or threaten a dog, or approach a dog that's sleeping or eating.
- Never leave children under 5 years of age alone with a dog and don't allow children this age to walk a dog.
- Don't grab dogs around the neck or tail.
- Introduce a child to a dog slowly and under close supervision.

If a dog does attack:

- Stand totally still and cover your face with your hands. Don't try to pat the dog — wait for the dog to lose interest in you. Keep your eye on the dog, but don't stare, because this can appear threatening to a dog.
- If you're knocked to the ground by a dog, don't attempt to get up; roll into the foetal position and lie as still as possible. (Trying to get up only encourages the dog to pull you back down, resulting in a more vicious attack.)

To treat minor bite wounds, carefully wash the area with water and antiseptic or soap. Apply antiseptic ointment and cover with a dressing, then apply a cold pack to reduce swelling. Check the wound daily for signs of infection (see Appendix A). Note that bites on the face usually require medical attention to avoid infection and scarring.

For severe bites, stop the bleeding by applying pressure to the wound with a clean cloth, then call an ambulance.

Dive-bombers

Whether you're walking in your own garden, to the local store or through a park, be on the lookout for swooping birds, especially during springtime.

The birds that are most likely to swoop on people in urban areas are the Australian Magpie and the Magpie-lark.

To find out about other Australian birds that can harm you and could venture into the suburbs, refer to Chapter 8. For information about beach birds, check out Chapter 10.

Magpie mayhem

The Australian Magpie is both loved and loathed in communities. Some people love them, feed them and encourage them, and even try to pet them. Others loathe them, because they're mighty defenders of their territory. They dive-bomb anything they perceive as a threat to their nesting site or feeding territory, and humans are most often the target.

Magpies, affectionately called maggies, are found throughout Australia, except in desert regions. They feature bold black-and-white markings and grow to a length of between 34 and 44 centimetres (13 and 17 inches). They're members of the crow family and are handsome rather than pretty. Magpies are sleek and sturdily built, and can be downright cheeky, begging and stealing food. Magpies are also talented song birds, producing an admiring and melodic call. They have even been known to mimic a human voice.

Defenders of the realm

Normally, Magpies are harmless, but they aggressively defend their territory while they're nesting or when adolescent males are being ousted from the family. This usually happens in spring between August and November in Australia. If they believe that you might threaten their nest, Magpies swoop down close, aiming for your head. Their beaks are powerful and can do considerable damage, causing cuts and abrasions or eye damage. They target both pedestrians and cyclists and can cause further injuries, such as bone fractures, if you're knocked to the ground or fall off your bike.

Australian Magpies are protected by law. This means you're not allowed to keep them as pets or harm them. If swooping Magpies become a danger in a particular area, contact the local council or the police.

Street smarts, Magpie-style

Magpies are extremely intelligent birds. They can recognise individual faces. If you ever threaten their nest or eggs, they will target you again and again in the future. But if you cross their path on a regular basis and have never threatened them before, they're likely to leave you alone.

These birds can even tell the time of day. For example, we've seen a family of Magpies sit on a railing outside a butcher's window waiting to be fed. If the butcher is a little late putting out meat scraps, the birds start pecking on his window as a gentle reminder.

Avoiding a Magpie attack

Most Australians are likely to be dive-bombed by a Magpie at least once. But you can take some measures to avoid being injured by a swooping Magpie. During the breeding season, between August and November, follow these guidelines:

- Don't approach or interfere with Magpie nests or eggs.

- If you're cycling (or riding a horse), wear a helmet. Some people paint a pair of eyes on the back of their helmet to frighten off the swooping bird. Get off and walk if dive-bombed.

- If you're walking, wear a hat and sunglasses. Some people wear an extra pair of sunglasses backwards or on top of their head — this may look a little strange but can deter a dive-bomber.

- Hold an umbrella or stick above your head while walking; this interferes with the Magpie's flight path.

- Don't panic and run; you're better to look confidently at the Magpie and keep walking — but get ready to protect your eyes, if necessary.

- If you're swooped at the same location more than once, try taking a different route for a few weeks.

- Walk or ride in a group: Magpies are more likely to attack an individual.

A Magpie impostor: The Magpie-lark

Magpie-larks look like small Magpies and are found in urban areas in most parts of Australia. Magpie-larks have more white markings on their bellies than Magpies. Despite their name and resemblance to the Australian Magpie, they're not closely related to their namesake. But they do swoop like Australian Magpies during breeding season (August to November) and are capable of causing serious injuries.

To avoid being swooped by a Magpie-lark, refer to the guidelines listed in the preceding section.

Treating beak injuries

For cuts and abrasions to the head caused by the Magpie or the Magpie-lark (and other swooping birds), first stem the blood flow by applying pressure to the wound with any type of fabric you have at hand, then wash the wound and apply antiseptic lotion. Applying a cold pack to your head can also help reduce swelling caused by the bird's beak striking your skull. A particularly bad gash of 3 centimetres (1 inch) or more may need stitching, particularly if the bleeding doesn't subside.

For eye injuries or suspected bone fractures, seek medical assistance immediately.

Chapter 15

Unwelcome Gatecrashers

*S*uburban homes provide an attractive temporary dwelling for some Australian bush creatures. You just never know when an uninvited and unwelcome guest might drop in for a visit. Some of them enjoy the living conditions so much that they even take up permanent residence.

In this chapter, we show you how to deal with most of the unwelcome visitors that invade properties in urban areas and cities. Some are native to this country, such as snakes, but most are foreign species — creatures from other countries that prefer to hang around people and homes rather than live in the Australian bush.

Snakes in the Garden

Brown snakes, tiger snakes, copperheads, death adders, pythons and even taipans are sometimes attracted to gardens and backyards. All snakes like to bask in the sun, in quiet places — doing so helps them maintain their body temperature. And the smell of some domestic animals, such as chickens (chooks), or pet mice, rats and birds, can lure them in, offering a free feed.

For information on Australian snakes and descriptions of the most common species, refer to Chapter 5.

Favourite haunts

Snakes hide out around chicken coops, compost bins, aviaries, wood-piles and under metal sheeting — anywhere that provides shelter and is near to food. Gardens close to native bush or creeks are more subject to snake visits than city dwellings, but many cities have large native gardens or reserves of land, which are attractive haunts for these slithering visitors.

To keep snakes away from your home, avoid leaving untidy piles of wood, other building materials or rocks lying around the house or garden. Neat piles are less likely to attract snakes. Keep grass short.

Dealing with an intruder

Be alert, not alarmed! Treat snakes with respect and keep your distance — they usually slither away.

If you hear your dog barking frantically in the backyard, there's a good chance it's cornered an animal from the bush that's lost its way. If the intruder is a snake, all the noise and attention from your pet will aggravate the snake, so you're best to call the dog off (if you can), go back inside and phone for an expert. Local councils usually have a list of trained 'snake busters', capable of collecting snakes, wild bees, possums and other creatures safely, and returning them to their natural habitat. Vets can assist if your pet is bitten.

Occasionally a snake may even venture inside a house. If you find a snake indoors, try to confine it to a single room by closing doors, then phone your local council.

If you're confronted by a snake, remain as still as possible. If the snake doesn't move off, back away very slowly. Try not to make any quick movements and don't move towards the snake. Most Australian snakes are protected by law, so don't attempt to catch or kill one.

If you get bitten, try to get a good description of the snake, because antivenoms are available for most species, then follow the first aid procedures described in Chapter 5 and Appendix A.

Ravenous Rodents

Beatrix Potter, the English author and creator of Peter Rabbit and other animal characters, may have had a fondness for rats and mice, but most people don't. Australia, like every other continent on earth (except Antarctica), has plenty of these rodents.

 Rats and mice are rodents, a group of animals that gnaw or nibble. Other animals in this group are squirrels and beavers.

Rats, rats, rats

Rats like living alongside humans, feeding off food scraps left in dumpsters, in garbage depots or in backyard bins. They nest wherever they can, as long as it's close to a food source. In bigger towns and cities, they nest in drains and sewers — places where they pick up disease-carrying organisms (like bacteria). Some of these diseases can be transmitted to humans by a bite, or by fleas that bite rats and then bite you. Rats are known to carry over 30 diseases harmful to humans, including the bubonic plague (for more information see 'Ferocious fleas' later in this chapter), endemic typhus, rat fever and infectious jaundice.

The rats that are considered dangerous in Australia are European. The common Black Rat (see Figure 15-1) and the Brown Rat arrived with early settlers in the 1780s. Native rats settled here about a million years ago and dwell in bush areas rather than around houses (see related sidebar). Native rats tend to stay away from humans and aren't dangerous.

 Distinguishing native rats from European immigrants isn't always easy. Black Rats (though not always black; some are shades of grey and brown) and Brown Rats have erect ears and long tapered tails. Native rats have more rounded features and are very shy.

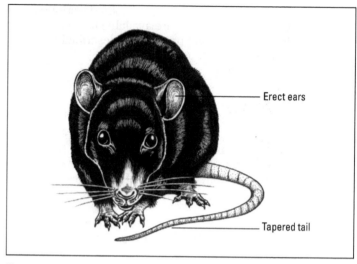

Figure 15-1: The introduced Black Rat.

Never corner a rat; it may attack you aggressively, scratching and biting. If you sight a loner, use a rat trap and some tasty treat to lure the rat in. If your home is infested with rats, call a professional rat-catcher.

Seed eaters

Mice, like rats, are disease carriers. They eat seeds and cause big headaches — economically speaking — for Australian cereal-crop growers. Also, like rats, mice like an easy feed, which householders often provide — boxed pantry food, fruit bowls, floor crumbs or food scraps in non-lidded rubbish bins . . . they eat almost anything.

The natives

Present-day native species of rats and mice are thought to be relatives of rodents that arrived on pieces of driftwood from Indonesia some six million years ago or made it overland from New Guinea when Australia and New Guinea were attached to each other, about one million years ago.

Discouraging rats and mice

Good hygiene is the best way to prevent rats and mice living near you. Wrap your rubbish and put it in a securely lidded bin. When travelling or camping, ensure your food scraps are burnt or buried deep, a good distance from the camp site. Store food securely in cans or jars — rats and mice can nibble their way through cardboard and plastic. And don't leave soap lying around — rats and mice love to gnaw on the fats in soap.

If you're bitten by a rat or a mouse, thoroughly cleanse the wound using an antiseptic, then check the site over the next few days for signs of infection (redness, swelling, weeping from the wound, pus, increased pain). A doctor may suggest a tetanus injection if the wound is nasty.

Recent Irritating Immigrants

Australia has its fair share of dangerous native species, and they don't always stay in their native surroundings. But some nasty introduced species live in Australia as well, such as Cane Toads, Fire Ants and wasps. All these creatures are recent arrivals — introduced to this country within the past 100 years.

Cane Toads: Introducing the unwanted amphibian

Australia has no native toads. Plenty of frogs are indigenous to this country, but not a single toad lived here until 1935. In that year, Cane Toads were imported from Hawaii to northern Queensland for breeding, in an attempt to kill off cane beetles that were destroying Australia's sugarcane crops and threatening the industry. Since then, the Cane Toad, native to South America, has become a menace — to local habitats, creatures and people.

Well, it seemed like a good idea at the time . . .

Cane Toads breed quickly and by 1937 more than 60,000 had been released into sugarcane crops along the northern Queensland coast to tackle the beetle problem.

Unfortunately, the introduction was a complete failure, because scientists didn't realise until after releasing the Cane Toads that their short runty legs didn't allow them to jump high enough to reach adult cane beetles on sugarcane, and that their breeding season didn't coincide with the ground-dwelling larval stage of the beetle.

The toads, however, found Australia ideal. Since their release, they've spread from northern Queensland across the Top End of Australia towards Western Australia, and southwards down the east coast towards Sydney. Figure 15-2 shows their current distribution.

If looks could kill . . .

Cane Toads are easy to recognise because they're ugly and heavily built — see Figure 15-3. They grow to 15 centimetres (6 inches) long, have dry green-brown or reddish-brown skin that's rough or warty looking, and pale mottled underbellies. Their eyes are well protected with bony ridges, and they have large swellings — containing venom glands — behind each eardrum. Their short hind legs have webbing between the toes, but their front legs don't.

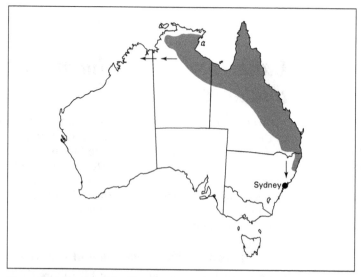

Figure 15-2: The Cane Toad is spreading westwards and southwards at an alarming rate.

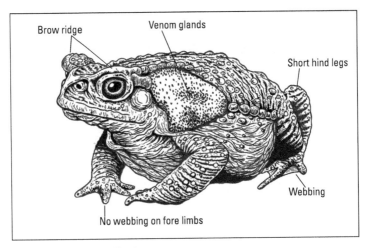

Figure 15-3: The Cane Toad — not cute.

The backyard bandit

Cane Toads can thrive in a range of habitats, from coastal regions to the edges of rainforests and in mangroves. However, they prefer open clearings, grasslands, woodlands and, especially, urban backyards. Cane Toads have ferocious appetites. The exterior lighting of houses and other buildings attracts a night-time supply of moths and other insects — which in turn attract Cane Toads. And people also attract Cane Toads when they leave food for pets outside. Yes, one of their favourite foods is dog food — dry or canned!

Pet dogs and cats, when first introduced to Cane Toads, often attempt to check them out, then get sick — read on to find out why. Take your pet to the vet if Cane Toads are in the area and your pet suddenly becomes sick and sullen. Usually, your pet will recover within a day or two (and won't ever attempt to touch a Cane Toad again!).

Glands that gush

A threatened Cane Toad exudes a white venom from glands around its neck (refer to Figure 15-3). It's capable of squirting a fine spray of the venom towards an attacker. The venom causes intense pain, temporary blindness and inflammation if it comes in contact with the eyes, mouth or nose.

Beating the venom

Some native animals have figured out how to eat Cane Toads without dying from the toxin. They kill and eat the toad in one of two ways:

- Flipping the toad onto its back and eating the internal organs, leaving the skin and the venom glands behind — some bird species do this.

- Only eating the toad's legs — a native rat species solution.

Scientists have also discovered that an Australian snake, the Keelback, may be immune to the toad's toxin.

Also, it appears that Estuarine Crocodiles, freshwater turtles and crayfish can eat small numbers of Cane Toads and remain unharmed. Wolf spiders can live off Cane Toad tadpoles.

Even better news may be on the horizon: Australian scientists, seeking a solution to the Cane Toad problem that doesn't harm native animals, have discovered that an Australian beetle (the Lavender Beetle), which native frogs ignore, kills Cane Toads when they eat them. Ah, the irony of it all!

Cane Toad culling has almost become a national night-time sport in some regions of northern Australia. If you're tempted to join in, wear gloves and thoroughly wash your hands after handling any toads.

If a Cane Toad squirts at your face, douse your eyes, mouth and nose with lots and lots of water. Do the same for any other part of your body that comes in contact with Cane Toad venom. Seek medical attention if symptoms persist.

Tracing the eco-disaster

What makes Cane Toads so horrifying is that they compete with native species for food and habitat — and win. They can reproduce more quickly, they're bigger and tougher at fighting diseases than native species, they have few predators, and they're pesticide-resistant. They also eat native mice, frogs and lizards, and are endangering Australia's honey industry by eating large numbers of Honey Bees. In fact, Cane Toads are one of Australia's worst environmental disasters.

To make matters worse, Cane Toads are poisonous to most animals that eat them. Snakes, lizards, Dingoes, Freshwater

Crocodiles and many pets have died after eating Cane Toads. Even Cane Toad tadpoles are poisonous to native fish. Although no human fatalities have been recorded in Australia, some people have died after eating the Cane Toad in South America.

Fire Ants: Fiery and feisty

Fire Ants, shown in Figure 15-4, are fiery little creatures. They may look similar to an ordinary house or garden ant, but they're smaller, ranging from 2 to 6 millimetres (0.1 to 0.2 inches) long, and copper coloured. They're extremely feisty, particularly if you disturb their nest.

Fire Ants swarm on you and use their tails to sting you again and again. Each sting injects a small amount of venom into your skin. A Fire Ant attack usually results in hundreds of painful stings. Within a few seconds, your skin feels like it's on fire — hence this creature's name — and after a few hours, watery blisters appear at the site of each sting. As the blisters heal they become terribly itchy. Scratching can break the skin, allowing infection to set in.

Figure 15-4: The Fire Ant: The sting is in the tail.

Use a cold pack to relieve the swelling and pain of stings. Wash with soap and water and leave the blisters intact — breaking the skin can lead to infections. People allergic to insect stings should seek medical attention immediately. On rare occasions, Fire Ant stings can cause a severe acute allergic reaction, resulting in life-threatening breathing difficulties and low blood pressure (anaphylaxis). See Appendix A for dealing with an emergency until help arrives.

Nesting in

Fire Ants seem to like building nests in open areas — in lawns and pastures, and along roadsides. Fire Ants build dome-shaped mounds as high as 40 centimetres (16 inches). Nests may be located beside logs, rocks or discarded timber. If you find a nest, don't touch it; stay clear and inform the local council.

Arrival and eradication plans

Fire Ants first came to Australia in 2001 aboard container ships from their native South America, hidden in timber and wood products. They're mostly confined to areas near Queensland ports, but colonies have also been found in Melbourne ports.

If Fire Ants aren't controlled they could spread throughout Australia, damaging crops and threatening native animals. Fire Ants are thought to be Australia's greatest ecological threat since the Cane Toad. The threat is so serious that, in Queensland, landowners discovering Fire Ants on their properties are legally obliged to inform the Department of Primary Industries and Fisheries, to ensure nests are destroyed and the pest's movements can be tracked.

Foreign wasps

Pretty? Maybe. Dangerous? Definitely. You can distinguish the European Wasp and the English Wasp from native varieties (covered in Chapter 14) by their striking yellow and black markings. Each yellow band has a pair of black spots on it. They're also larger than native wasps.

The European hitchhiker

The European Wasp (see Figure 15-5), native to Europe and low-lying regions of Asia and North Africa, is a worldwide traveller, having hitched rides to Australia in cargos of fruit. European Wasps were first discovered in Tasmania in 1959, but within 20 years had established themselves in many parts of Australia. Their numbers are now beyond eradication.

The European Wasp is a stout creature, measuring some 12–15 millimetres (0.5–0.6 inches) long, with black antennae, two pairs of wings and a distinctive flying style — curling its legs close its body when in the air.

The distribution of European Wasps is mostly limited to Australia's cooler and wetter regions:

- Coastal southern Australia (including Tasmania)
- The cooler country areas of New South Wales
- Around the hills of Adelaide, South Australia
- From Perth to Albany in Western Australia

The English hitchhiker

The English Wasp is a close relative of the European Wasp and has similar markings (see Figure 15-5). It has been discovered in the eastern suburbs of Melbourne, in Gippsland (Eastern Victoria) and in Tasmania. Treat English Wasps as you would European Wasps: With caution, and follow the same first aid treatment if you're stung.

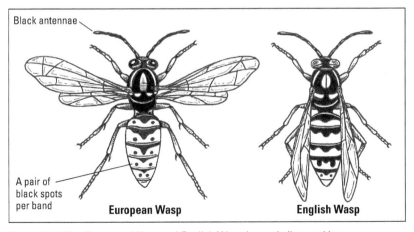

Black antennae

A pair of black spots per band

European Wasp　　　　**English Wasp**

Figure 15-5: The European Wasp and English Wasp have similar markings.

Following the scent

Unlike bees (refer to Chapter 14), wasps can sting several times, inflicting painful swellings. And the smell of meat, sweet food or drink lures worker wasps from their nests. At the height of summer, these wasps often crawl into drink cans, inflicting the nastiest stings of all — inside your throat — when you take a sip from the can. If a wasp is swallowed, that painful swelling can lead to breathing difficulties, particularly in young children. Drinking through a straw averts such stings.

Use a cold pack to help relieve the pain and swelling associated with stings. Severe cases, such as allergic reaction or a sting to the back of the throat, can be life-threatening: Be prepared to give mouth-to-mouth resuscitation and cardiac massage, while someone calls an ambulance. See Appendix A for more details.

Communal living

European Wasps and English Wasps form large communal nests underground or in cavities in walls or in trees. They build their nests, made of chewed wood fibres, with only a small entrance hole. European Wasp nests are grey, whereas English Wasp nests display a wave-like pattern of various shades of brown. This is because the European Wasp uses chewed-up dry and weathered wood to make its nest, whereas the English Wasp selects rotted wood.

More aggressive than bees, these wasps attack when their nest is disturbed — poking around their nests with sticks or other tools can send them into a frenzy. Call a pest control company if you discover a nest in your garden.

Here are some things you can do to ensure you're not encouraging European Wasps or English Wasps to gatecrash your outdoor activities at home:

- Keep lids on rubbish bins closed
- Avoid leaving pet food outdoors
- Cover food for as long as possible while barbecuing
- Don't leave fallen fruit or food scraps outside
- Keep compost heaps covered
- Cover swimming pools when they're not in use

Tiny Bloodsuckers

Some pests, such as lice, bedbugs and fleas, are found worldwide. They feed on your blood, causing swelling and itchiness. Some of them spread one of the most deadly diseases ever known — the horrid Black Plague! But they're all tiny, tiny, tiny — less than 8 millimetres (0.3 inches) long.

Lice aren't nice

More than 3,000 species of lice exist worldwide. These external parasites live on mammals and birds, feeding on dead skin or feathers, the skin's oily secretions and their host's blood. They can spend their whole life on the one host animal.

Each species is highly specialised, feeding only on certain areas of their host's body. Three species commonly infest humans.

Head Lice

Head Lice are bloodsucking human parasites. As their name suggests, they're found on the head of their host, preferring areas behind the ears or at the back of the head.

Head Lice can only survive a few days without a human host — they can hide in, but can't reproduce in or infest furniture, bedding or pet areas. Transfer of Head Lice is by head-to-head contact or by sharing brushes and combs, ribbons or hair bands, hats, pillows, blankets and doonas.

The first signs of an infestation are an itchy scalp or constant head scratching. A female Head Louse can lay up to 300 eggs, called nits. The hatchlings will eat 3,000 to 6,000 blood meals in a month — itch, itch, scratch, scratch!

Head Lice are very small (2.5 to 3.5 millimetres or about 0.1 inch in length) with a light brown to reddish-brown flattened body, well-developed eyes and six legs. The claws on each leg help the lice cling to hairs, making them hard to brush out.

Constant scratching can lead to skin lesions and infections and, in some cases, even swollen lymph glands. But these lice are usually more of a nuisance than dangerous; they can disrupt a person's sleep and concentration, making them irritable and ill-tempered, but the lice don't spread diseases of any kind.

To identify a Head Lice infestation, look for:

✔ Eggs or nits attached to the base of hair shafts or adult lice scurrying through the hair

✔ Dirt (faeces or old skin castes of the lice) on pillows or blankets

✔ A strange smell — left untreated, an infestation causes the hair to mat and give off a putrid odour

Pubic lice

Pubic Lice — sometimes called Crab Lice or 'crabs' — also feed on human blood. The name, *crabs*, comes from their crab-like appearance. Pubic Lice are grey, small (1–3 millimetres or about 0.1 inch) with an oval body shape and tiny head. Their front legs have large claws, whereas their back legs have only minor claws. The claws allow a Pubic Louse to take hold of a coarse hair while they pierce the skin and suck their meal of blood.

Pubic Lice are slow-moving insects that are found on human pubic and other coarse hair areas, like moustaches and beards, eyelashes, armpits, and sometimes chest and abdominal hair. Occasionally, Pubic Lice can be found around the head's hairline. They travel from one person to another by sexual contact or sharing beds (or in a bed recently vacated by someone), clothing or towels. Pubic Lice aren't known to carry diseases.

A Pubic Louse bite looks like a small raised red spot surrounded by a swollen area and is extremely itchy. After a day or two the bite takes on a grey-blue colour. If the infestation involves eyelashes, the eyelids become red and swollen. Scratching can lead to skin infections, and swollen lymph glands.

Body Lice

Human Body Lice are a bloodsucking species of lice that live only on clothed areas of the human body. They spread from person to person when people huddle together or share bedding and furniture.

Adult lice are 2 to 4 millimetres (0.1 to 0.2 inches) in length, grey in colour, with flat, long oval abdomens, a distinct head and small eyes. Each of their legs ends in a claw, which is used to grip cloth fibres and body hairs.

Human Body Lice feed in soft skin folds; females lay their eggs (or nits) along the seams and hemline of clothing in close contact to the skin. The Lice scurry away and hide in a seam or a crease when disturbed.

Human Body Lice cause intense itching. Bites present as tiny red spots with a raised centre. An ongoing infestation can cause constant itchiness and irritability, headache, joint pain, fever, tiredness, loss of appetite and a fine body rash, made up of tiny red dots.

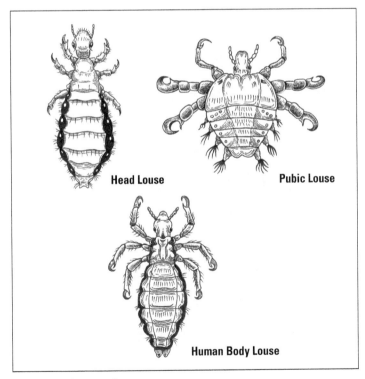

Head Louse　　　　　**Pubic Louse**

Human Body Louse

Figure 15-6: Common lice.

All aboard!

Head Lice, Pubic Lice and Human Body Lice spread easily from person to person, so if detected they need to be treated immediately. Personal belongings, clothing, blankets and furnishings also need to be treated.

Pharmacies carry shampoos or similar products that are designed to kill Head Lice and Pubic Lice. No prescription is necessary. Two applications, seven days apart, are necessary to ensure all adult and developing nits are killed. To remove unhatched nits or eggs, use a fine-toothed comb. Treat eyelash and eyebrow infestations by applying petroleum jelly twice a day for 7–10 days.

You need a body wash (prescribed by a doctor) to kill Human Body Lice. Apply the wash thoroughly over your body and allowed it to dry. Repeat for three days to eradicate the problem.

You need to treat personal and bedding items to ensure further infestation doesn't occur. Soak combs, brushes, hair bands and other personal items in an antiseptic solution or lice-treatment product. Wash pillows, bedding and bed toys (like teddy bears) in hot water and then dry in a clothes dryer on the hot cycle for 20 minutes or in full sun. Also, examine (and treat, if necessary) all other occupants in the household.

Don't let the bedbugs bite

Bedbugs are found worldwide because they stow away in people's luggage and clothing. They're wingless insects, red-brown in colour — redder after a meal of blood.

The adults are 4–5 millimetres (about 0.2 inches) long, oval and disc-shaped. Bedbugs are fast movers if light is shone on them. They hide in dark narrow cracks and crevices close to where people sleep — in the seams of mattresses, the cracks between floorboards, the frames of paintings and in carpets, making them hard to find. They're seldom found on the human body itself.

Bedbugs tend to congregate in groups and infestations are often associated with a distinctive sweet, sickly smell. Another sign of their presence is tiny blood spots on mattresses, bedding and other furnishings.

How to catch bedbugs

Because bedbugs shun sunlight, you're best to use a torch, in the dark, to try to find them. Or, you can stick double-sided tape along the seams of a mattress and after a day or two check to see what you've caught. If you discover bedbugs you need to fumigate your home to get rid of them.

Pest controllers use synthetic pesticides to eradicate bedbug infestations. Bedbugs can travel and hide some 30 metres from their victims, so neighbouring rooms also need to be treated. Clothes need to be washed in hot water and dried in the sun or on the hot cycle of the clothes dryer. Delicates can be wrapped and placed in the freezer.

Non-chemical techniques using black plastic and sunlight or blasting an item with hot air may work for small items, but isn't known to be successful on bigger articles. Good housekeeping practices and reducing the number of cracks and crevices in a home discourages infestations.

The Common Bedbug (shown in Figure 15-7) and the Tropical Bedbug are the two most commonly found species of bedbugs in Australia. They're very similar in appearance, but the Tropical Bedbug is a little slimmer. They both produce similar symptoms when they bite and treatment is identical.

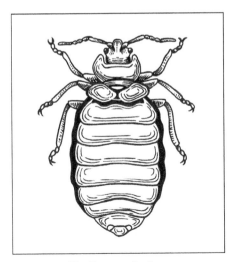

Figure 15-7: The Common Bedbug.

Heat-seekers

Bedbugs locate humans by the warmth of their bodies and the carbon dioxide they emit when breathing. And because they shun light, they tend to bite you while you're asleep in bed — they probably wouldn't be called bedbugs otherwise.

They're well-designed to bite human bodies and suck blood, having piercing mouthparts and saliva that prevents the victim's blood from clotting. During the day they lay quiet, digesting their blood meal of the night before. However, bedbugs will bite during the day if they're hungry enough. And they can survive for long periods of time without feeding.

Dealing with bedbug bites

A bedbug can get its fill from one bite, but if you move during the process you may end up with a line of small bites on your body. This distinctive pattern — say, three bites in a line — is sometimes referred to as 'breakfast, lunch and dinner' and is often found along blood vessels lying close to the skin's surface.

Although bedbugs have been known to carry disease-causing organisms, such as Hepatitis B and plague, they're not known to transmit these to humans. For most victims, bedbug bites cause mild irritation. The real danger lies in infections and allergies. Scratching the bites can break the skin and allow infection and subsequent scarring. Some people develop allergic reactions to bedbug bites and, after being bitten, suffer *delusional parasitosis* (a belief that the victim is infested with parasites) and post-traumatic stress.

Wash the bite with an antiseptic lotion and watch for signs of infection (redness, swelling, weeping from the wound, pus, increased pain) over the next few days. If allergic symptoms occur, seek medical advice.

Ferocious fleas

Fleas are wingless, bloodsucking insects that live on mammals (including humans) and birds. If you've ever owned a dog or a cat, you probably already know plenty about them!

The warm, hair-covered skin of dogs and cats provides the perfect place for females to lay their eggs. The eggs are scratched or fall off and hatch into legless maggots, which can crawl into warm places like carpet, kennels, baskets and bedding. They feed on anything organic they can find, including hair, dead skin and dried blood. From a few weeks to a few months later — depending on temperature, humidity and food availability — an adult flea emerges.

A host of diseases

Fleas are capable of transmitting parasites and diseases from their animal hosts to humans. Dog Fleas and Cat Fleas, for example, can carry a tapeworm from dogs or cats to humans. Rat Fleas can carry more serious diseases to humans, including murine typhus and the age-old killer, the Black Death (see related sidebar). Symptoms of murine typhus include headache, very high fever, a red rash, nausea, vomiting, abdominal pain, aching joints and a dry cough. Despite the nasty symptoms, though, murine typhus is rarely fatal.

A *parasite* is an animal or plant that gains its nutrition from another plant or animal rather than finding its own food. Some parasites, such as tapeworms, are internal and live inside an animal. Other parasites latch onto the outer skin of an animal. Fleas, ticks, lice, bedbugs and leeches all fit this category.

Pinpointing (and controlling) the trouble-makers

About 30 species of fleas are native to Australia, but only the intruders from other countries feed on human blood. Among these irritating immigrants are the Cat Flea, Dog Flea and Rat Flea. Of these, the Cat Flea, shown in Figure 15-8, is the most common biter of humans. And don't be fooled by its name — Cat Fleas live on both cats and dogs.

When fleas dine on *your* blood, they often leave clusters of bites. Their saliva affects your skin, causing swelling and irritation. Some people are more sensitive to flea bites than others.

Applying a soothing antiseptic cream is usually all you need to do to relieve bite symptoms. Don't scratch the affected area. If irritation is severe or continues for more than two or three days, seek medical treatment.

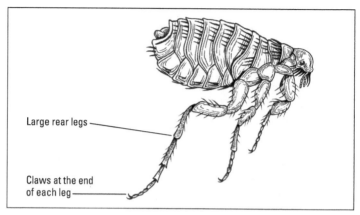

Large rear legs

Claws at the end
of each leg

Figure 15-8: An adult Cat Flea ranges in colour from light to dark brown.

The key to ridding your home of fleas is eliminating suitable breeding places. For most people this means treating your pet dog or cat (check with a vet for the best products) and washing your pet's bedding often. Keeping your home environment clean is the best way to keep rats and mice — the other likely flea hosts — away. For tips, refer to 'Discouraging rats and mice' earlier in this chapter.

The Black Death

Fleas are responsible for more human deaths than any other creature on the planet. During the 14th century about 25 million people died from the Black Death, a disease that took on three deadly forms — bubonic plague, pneumonic plague and septicaemic plague. The disease became known as the Black Death because large black bruises appeared on the bodies of victims before they died.

The Black Death originally only affected rats and other rodents. But as rodents died from the disease, the fleas that had been drinking their blood required a new host — humans — transmitting the deadly disease to people as well.

Further major outbreaks of the Black Death among human populations killed about 100,000 people in 1565 during the Great Plague of London and about 12 million people in Asia at the end of the 19th century. A vaccine for the Black Death is now available, but the disease still causes about 1,000 deaths each year.

Part V
The Part of Tens

Glenn Lumsden

Insects can test anyone's patience.

In this part . . .

We couldn't resist making a list of the ten deadliest Australian creatures. Of course, such a list changes with time, as new antivenoms and medical treatments are developed.

We also put together ten of the most effective ways to stay safe from dangerous creatures, and list ten places where you can check out Australia's deadliest creatures, up close.

Chapter 16

Australia's Ten Deadliest Creatures

*W*hich creatures are Australia's deadliest? It depends on what you mean by *deadly*. If you judge deadly by what you read on the front pages of newspapers, you'd probably say sharks or crocodiles. Their attacks attract media attention because they're ferocious and can cause massive injuries and blood loss. But some of the deadliest creatures attack without much drama, injecting venom that can quietly kill a human within a few minutes. Attacks by these creatures tend to attract less media attention.

We've based our selection of deadliest creatures on the likelihood of human fatalities. This doesn't only depend on the toxicity of the venom an animal can inject, or how much damage sharp and jagged teeth can inflict. We also take into consideration how common these creatures are in Australia, and how close they live to human populations. Nearly all the deadly creatures that make this chapter also feature in the colour photo section in this book — check 'em out!

Snakes

Snakes are the deadliest Australian creatures. In fact, the ten most venomous snakes in the world, according to the toxicity of their venom, are all found in Australia.

The Inland Taipan (also called the Fierce Snake) has the most toxic venom of all snakes — about 50 times the killing power of the more well-known Indian Cobra, which doesn't even make the top ten world list. Yet the Inland Taipan hasn't been known to cause human deaths, for two reasons: It lives in remote areas and has small fangs, so it can only inject a tiny amount of its toxic cocktail when it does strike. Other Australian snakes high in the top ten world list of venomous snakes include the Eastern Brown, the Coastal Taipan and the Eastern (or Mainland) Tiger Snake. These snakes have all been responsible for human deaths.

Snakes have killed an average of two people each year over the past 30 years. If it weren't for scientific research and the development of antivenoms, the toll would be much greater. Most snakes retreat when humans approach and only attack when provoked or cornered — whether deliberately or accidentally.

Although only a quarter of Australia's snake species are dangerous, they're difficult to identify, so assume that any snake you meet in the wild is dangerous. For more information on identifying snakes, refer to Chapter 5.

Honey Bees

Honey Bees rank second to snakes as Australia's deadliest creatures. In a continent that's home to snakes, sharks, crocodiles and the Australian Box Jellyfish — the world's deadliest jellyfish — this fact may surprise you.

What makes Honey Bees so dangerous is that they're found in large numbers where humans tend to spend a lot of time — in suburban backyards, parks and gardens. Another reason they're graded as the number 2 killer is that many humans are allergic to bee stings. Allergic reactions vary from minor irritations and rashes, to swelling, stomach cramps, vomiting and diarrhoea. A severe allergic reaction can lead to death. For more information about Honey Bees, refer to Chapter 14.

Dogs

Sometimes the greatest threats to your health are much closer than you think. On average, pet dogs (covered in Chapter 14), are responsible for one death per year. Sadly, the victims of these fatal attacks are often children or elderly people. And sometimes the offending dog has been a loved and trusted pet for many years. Wild Dingoes have also been known to kill children (refer to Chapter 7).

Most of the thousands of dog bites in Australia each year occur at home. Children are more vulnerable to attacks because they're more likely to accidentally provoke a dog. Also, because children are smaller, their heads and faces are closer to the dog's mouth, so attacks are likely to cause more serious injuries.

Most dog attacks on children are preventable. Infants and toddlers should never be left alone with a dog, no matter how small or well-trusted the animal is. Older children must never provoke a dog — deliberately or accidentally.

Sharks

Sharks have been portrayed by the movie industry as killing machines with an appetite for human flesh. This is far from the truth. Most species of shark are too small or timid and avoid humans. In fact, most shark attacks are cases of mistaken identity, when large sharks incorrectly identify humans as their usual prey.

About 200 fatal shark attacks have occurred in Australian waters since the first recorded fatality in 1791 — an average of a little less than one per year.

The Great White Shark, the Tiger Shark and the Bull Shark have all been involved in fatal attacks in coastal waters on swimmers, surfers and divers. Non-fatal attacks often result in horrific injuries. Some of the sharks found in the open ocean are capable of attacking and killing humans, too. The Oceanic Whitetip, for example, has been known to feed on survivors of shipwrecks and plane crashes. And the Short Fin Mako has attacked the occupants of fishing boats.

For lots more information on how sharks behave, check out Chapters 10 and 12.

Saltwater Crocodiles

Australian Saltwater Crocodiles (also known as Estuarine Crocodiles, or called 'salties' by locals) inhabit the rivers, estuaries and coastal waters in Australia's tropical north. They use their powerful jaws to crush their prey — bones and all. Big, full-grown salties prey on mammals as large as water buffalo, so for them, humans are considered food.

Saltwater Crocodiles are territorial and have no fear of humans. If you're careless enough to invade their space, you could be ferociously attacked.

Between 1971 and 2004, Saltwater Crocodiles launched at least 62 unprovoked attacks on humans. Of these, 17 were fatal. Although the number of unprovoked attacks is increasing at a remarkably rapid rate, the number of fatalities isn't. For example, from 1971 to 1980 only five croc attacks were recorded, four of which were fatal. Yet during the four-year period from 2001 to 2004, salties raked up 38 unprovoked attacks, five of which were fatal. The increase in the number of attacks is mainly due to the crocodile's growing population and the thriving tourism industry in crocodile country — refer to Chapter 4 to find out more about the Saltwater Crocodile's habitat and how to avoid an attack if you're visiting the tropics. Improvements in emergency services and medical treatment have prevented the death toll from rising at the same rate as the number of attacks.

The Australian Box Jellyfish

The Australian Box Jellyfish is the largest and deadliest species of a group of jellyfish that has a four-sided, box-shaped body. A sting from this jellyfish causes excruciating pain. The lethal venom in its long tentacles has killed at least 70 people in Australian waters since 1900.

The presence of this deadly creature in tropical coastal waters makes it too dangerous to swim in beautiful, remote tropical beaches during the wet season. However, at popular tourist beaches, stinger nets are set up to provide safe zones for swimmers.

Despite its size — about 30 centimetres (1 foot) across — the Australian Box Jellyfish is very difficult to see in the water, which makes it all the more dangerous.

For more information about the Australian Box Jellyfish, as well as other deadly stingers, refer to Chapter 10.

Wasps

Anybody who has been stung by a wasp knows how painful the experience can be. In most cases, the only other effect of a wasp sting is localised swelling. However, if you're allergic to wasp stings, the consequences can be fatal. Unlike bees, wasps can inflict multiple stings. And stings are common because homes and gardens provide ideal nesting places. The worst stings occur when a wasp nest is disturbed — large numbers of very angry wasps then attack, each capable of inflicting multiple stings.

Between 1979 and 1998, seven people died following wasp stings. All seven fatalities occurred in rural areas; all victims were male and all of them died within an hour of being stung, as a result of allergic reactions to wasp venom. Six of the victims were working outdoors or gardening when they were stung. The other was walking in a national park. It's believed that all of the fatal stings were inflicted by native paper wasps (for more information about paper wasps, and other native wasps, check out Chapter 14).

Despite its reputation, the introduced European Wasp (covered in Chapter 15) has not yet been attributed with causing any deaths in Australia. These wasps are potentially deadly, but they tend to nest in urban areas, so sting victims usually have quick access to emergency services.

Spiders

The only two spiders that have officially killed people in Australia are the Red-back Spider and the Sydney Funnel-web Spider. However, these venomous creatures are certainly not the killers they used to be.

Before the development of Red-back antivenom in 1956, at least 14 people died after being bitten by this spider. Since then, only one confirmed Red-back Spider bite fatality has been recorded. The big, black Sydney Funnel-web Spider has killed at least 13 people, but since the development of a successful antivenom in 1980, no one else has died.

Even though antivenoms are life-saving, you shouldn't be complacent around these spiders — especially in the case of the Sydney Funnel-web, whose venom acts quickly, especially in children. Both of these spiders are still deadly if enough venom is injected and the correct antivenom is not administered.

For more information on spiders, refer to Chapter 13.

Ants

Most of the ants you see around the home are less than 2 or 3 millimetres long (about 0.8 of an inch) and don't sting. But some larger species can be deadly to people who are allergic to their venom. Many people allergic to bee stings are also allergic to ant stings. At least seven people have died since 1979 as a result of allergic reactions to ant stings.

The killer ants in Australia are bull ants (of which there are many species) and the closely related Jumping Jack (also known as the Jumper Ant). Bull ants can be as big as 30 millimetres long, while Jumping Jacks are less than half that size — but still big compared with the tiny ants that might invade your kitchen. Bull ants and Jumping Jacks are very aggressive when disturbed and deliver very painful stings. To find out where they live and what their nests look like, refer to Chapter 14.

Ticks

It's hard to believe that something as small as a tick can be deadly. Since 1900, ticks have been responsible for killing more than 20 people in New South Wales alone. The most dangerous tick in Australia is the Paralysis Tick, which is confined to a 20 to 30 kilometre-wide strip along the east coast of Australia. For more information about the Paralysis Tick, check out Chapter 6.

Ticks latch on the skin of mammals, including humans, for days at a time and feed on their blood. As they feed, they leave behind saliva, which in the case of the Paralysis Tick, contains a paralysing toxin. In some cases paralysis is severe enough to cause death, especially if combined with an allergic reaction.

Chapter 17

Ten Ways to Stay Safe

Staying safe from Australia's dangerous creatures involves three critical factors:

✔ Avoiding encounters in the first place

✔ Ensuring that you know how to respond if you do encounter them

✔ Knowing what to do if you're attacked

In this chapter, we provide some advice that may help you stay safe. The topics we raise here are covered in more detail throughout this book.

Arm Yourself with Knowledge

Knowledge is a powerful weapon. Knowledge about animals, where they live, their behaviour and the correct first aid is invaluable for avoiding injury or surviving an attack. Remember, you don't have to leave your home to encounter a dangerous animal — like a wasp or a poisonous snake — so don't wait until you go on a trip to find out what to do.

Reading this book is a great place to start, but you can find heaps of information about animals on the Internet and in other books. Consider taking a first aid course. St John Ambulance Australia offers a range of first aid courses, including online courses. To find out more, visit the St John Ambulance's Web site at www.stjohn.org.au.

Avoid Dangerous Areas

You're best to stay away from some areas — unless you have a very good reason for being there and have the right protective clothing and equipment.

For example, you'd be unwise to swim or surf at beaches along the Great Australian Bight, because this area (refer to Figure 1-1) is frequented by the Great White Shark. Swimming in the sea in murky water or between dusk and dawn at any beach is risky — a shark might mistake you for its prey.

Whatever you do, don't swim in tropical beaches during the wet season, when the lethal Australian Box Jellyfish is about, except within the stinger nets erected at popular spots. Also, be wary of estuaries and river banks. These places are frequented by crocodiles, stingrays and spiky fish.

On land, stay out of long grass, especially during summer — long grass is a favourite haven for deadly snakes. And steer clear of ant nests, and bee and wasp hives, just in case you accidentally provoke a collective attack.

Heed Warning Signs

Local councils and other government authorities place warning signs in areas where wildlife is likely to be present. Signs on many beaches frequented by dangerous jellyfish, sharks and other creatures tell you when and where it's safe to swim, or when the beach is patrolled by lifeguards. Other warning signs highlight stretches of road likely to be crossed by kangaroos, koalas and wombats. Also, keep an eye out for signs in national parks and conservation reserves that request visitors not to disturb the environment.

Several fatal crocodile attacks have occurred when tourists have ignored Saltwater Crocodile warning signs in the tropics. Don't let this happen to you. *Always take heed of warning signs.*

When far away from home, talk to the locals. Striking up a conversation about the local wildlife is a great way to get to know people who live in the region, and they can also provide a wealth of information about what to see and do — and where dangerous creatures live and seasonal variations.

Be Prepared

When you're going on a trip or outing, make sure that you're prepared. If you're bushwalking, camping or taking a long journey by road, you need water, maps, a list of emergency contacts (preferably in a plastic bag), insect repellent, sunscreen and the appropriate clothing. Always take a torch and spare batteries if you're going to be out at night. And if you know that you're allergic to bee, wasp or ant stings, or if you're taking medication, make sure that you tell your companions.

In an emergency, a mobile (cell) phone is the easiest way to summon help. If you're planning to visit remote areas that are unlikely to have mobile phone coverage, consider renting or buying a satellite phone.

For more information about what to pack when you travel, refer to Chapter 3.

Don't Go Alone

It's always safer to go walking, riding, swimming, surfing and diving with company, rather than on your own. If you're stung or bitten by a dangerous creature you may have to rely on another person to assist you or to get help. Or, if you're stung by a lethal jellyfish, bitten by a brown snake or attacked by a Great White Shark, you may not be able to treat the wounds or reach help by yourself.

Keep Your Distance

All Australian dangerous creatures can be observed safely close-up in zoos, wildlife parks or aquariums — see Chapter 18 for our pick of the best — but if you're outdoors or in the bush, observe the animal from a distance. Getting too close could cost you your life.

Many dangerous creatures only attack if they're cornered, threatened or provoked. However, always keep your distance from nests, eggs or offspring. No matter how safe an animal may appear to be, most attack ferociously to protect their territory or young.

Wear Appropriate Clothing

If you're walking in the bush — whether inland or on the coast — always wear a sturdy pair of shoes or boots and long trousers. This also applies to wading in rivers or creeks — you never know what you might stand on or accidentally disturb.

Even when you're working in the garden, be prepared for encounters with spiders, snakes, ticks, wasps and other critters. Wearing a good pair of garden gloves could save you from a potentially lethal bite or sting.

Watch Where You Put Your Hands and Feet

Snakes, spiders, scorpions, spiky fish, octopuses, stingrays and many other dangerous creatures seek shelter or lie in wait for prey where you can't see them. If you put your bare hands or feet in places where they may be hiding, you're likely to come off second best.

On the bush trail, or in the garden, keep your hands out of hollow logs, cavities in rocks, ground cover, long grass or any other places where animals might be sheltering. When camping (or even indoors), always check your gear to make sure that a small and possibly dangerous creature hasn't snuck into your bedding or clothes.

When exploring rocky coastal shores, don't put your hands in rock pools. Some rock-pool creatures also hide in mud or sand in shallow waters, so watch where you put your feet, too.

Don't Feed the Animals

Wild or feral animals usually fear humans. But if you feed these creatures, they can lose their fear and aggressively seek more food. You also encourage unwanted visits from wild or feral animals if you leave food, food scraps or bait around your camp site. You may even end up being a tasty meal yourself if you prepare food or clean fish near the water's edge in the tropics, where crocodiles may be lurking.

Don't Panic

If you accidentally encounter or corner a dangerous creature, don't panic. You need to keep a clear head and think quickly about the sensible way to respond.

After an attack, the action that you or your companions must take depends on which creature has struck, and hopefully this book can help. If you don't panic, it's easier to identify the animal; and staying calm enables you to administer or gain the most from first aid.

Chapter 18

Ten Safe Places to See Australian Wildlife

*T*he very best way to meet Australia's fascinating animals is by visiting their true homes. Most live in their natural environments, in national parks, state forests, marine sanctuaries and many other conservation reserves across Australia. You can join an adventure tour to look for them, or do your own thing, but we can't guarantee that you'll see many of them. Of course, anticipation and searching is always worth the journey, but don't forget that when you enter an animal's territory, you may accidentally place yourself in danger.

Other places allow you to get up close and personal with a range of Australian creatures — without any nasty surprises. In this chapter, we list ten of our favourite places.

Australia Zoo

Made famous by the late Steve Irwin, Australia Zoo (www.australiazoo.com.au) on the Sunshine Coast in south Queensland is home to the Crocoseum, probably Australia's most famous Saltwater Crocodile enclosure, where you can watch these creatures' powerful jaws in action. Australia Zoo is also home to many other dangerous Australian animals, including brown snakes, taipans and death adders.

Australian Reptile Park

Australian Reptile Park (www.reptilepark.com.au) is less than an hour's drive north of Sydney. Given its name, you would think only reptiles live here, but at this park you can meet deadly and not-so-deadly spiders, scorpions, Dingoes, Platypuses, Southern Cassowaries and many other creatures. You can even watch a Sydney Funnel-web Spider being milked for its venom. Australian Reptile Park is the major Australian supplier of snake and spider antivenom.

Taronga Zoo

Like all metropolitan zoos in Australia, Sydney's Taronga Zoo (www.zoo.nsw.gov.au) houses a diverse range of animals. Taronga Zoo is home to crocodiles, venomous snakes, the Dingo, Southern Cassowary, Tasmanian Devil, kangaroos and many more. You can also see most of these animals in zoos in other cities, including Melbourne, Adelaide, Perth, Brisbane and Canberra, but the spectacular views over Sydney Harbour makes Taronga Zoo an extra special place to visit.

Healesville Sanctuary

On the outskirts of Melbourne, Healesville Sanctuary (www.zoo.org.au/healesville) is a great place to discover more about Australian wildlife, including some of the most dangerous species. Healesville Sanctuary provides an opportunity to meet and talk to the people who look after snakes, Tasmanian Devils, Platypuses, Koalas and kangaroos during its Meet the Keeper presentations. What's more, a Birds of Prey Show demonstrates the flight and hunting skills of birds like the Wedge-tailed Eagle.

The Aquarium of Western Australia (AQWA)

Marine aquariums in Australian cities offer you a great way to see marine creatures without getting your feet wet — or bitten off. Sydney Aquarium, Melbourne Aquarium and Underwater World at Mooloolaba on the Sunshine Coast, Queensland, are three worth visiting. Another you won't want to pass up if you're in Perth is the Aquarium of Western Australia (AQWA).

The Aquarium of Western Australia (www.aqwa.com.au), located at Hillarys Boat Harbour, is home to a variety of Australian marine creatures, including sharks, stingrays, jellyfish and several species of deadly fish. Here, not only can you get a really close look at marine creatures, but — if you dare — you can also swim with sharks — Grey Nurse Sharks, Wobbegongs and other smaller species. Among the deadly marine creatures in the tanks are Reef Stonefish, cone shells, sea snakes and blue-ringed octopuses.

ReefHQ

ReefHQ (www.reefhq.com.au) in Townsville in tropical north Queensland is amazing. As the world's largest living coral reef aquarium, it provides an insight into the marine life that inhabits the nearby Great Barrier Reef. In and around the coral you can see sea snakes, venomous spiky fish, sea urchins and the Crown of Thorns Starfish. At ReefHQ you can also see reef sharks, stingrays and a replica shipwreck through which all sorts of marine creatures swim.

Australian Museum

You may never get a better view of creatures such as bees, wasps, deadly spiders, centipedes, scorpions and ticks than at the Australian Museum (www.austmus.gov.au) in Sydney. Although they may not be alive, all the creatures are real and you can see them up close without being stung. The Australian Museum also displays specimens of birds, including the Southern Cassowary, Emu, Wedge-tailed Eagle, kookaburras and the Magpie. It even offers slightly spooky torchlight tours at night for 6–12 year olds.

Sea World

Smack bang in the middle of the Gold Coast in southern Queensland is Sea World (www.seaworld.com.au), an adventure–fun park that features the world's largest man-made lagoon for sharks, called Shark Bay. Here you can view sharks up close, from above and from below the water, through a window. Or you can swim with some of the smaller sharks or dolphins. If you're more adventurous, try being lowered in an acrylic cage to watch the larger sharks, including the deadly Tiger Shark, being fed. Sea World is also home to stingrays, sea lions, porcupine fish and sea urchins. Sea World has its own marine rescue team, which rescues and cares for whales, dolphins and turtles.

Territory Wildlife Park

Territory Wildlife Park (www.territorywildlifepark.com.au), about 45 minutes by car or bus south of Darwin, in the Northern Territory, is one of the few places where you can safely get a close look at feral pigs and water buffaloes. Territory Wildlife Park also houses other dangerous creatures that inhabit tropical ecosystems, including crocodiles, stingrays and venomous snakes.

Currumbin Wildlife Sanctuary

Currumbin Wildlife Sanctuary (www.currumbin-sanctuary.org.au), on Queensland's Gold Coast, began as a bird sanctuary in 1947, but has expanded to house a wide variety of Australian wildlife, including snakes, crocodiles and Tasmanian Devils. Currumbin Wildlife Sanctuary's shows and displays provide opportunities to get close to and find out about Australian creatures, including some of the dangerous ones. All of Currumbin's revenue is channelled into conservation, public education and the care of sick and injured wildlife.

Part VI
The Appendixes

Glenn Lumsden

The Australian Coat of Harms

In this part . . .

We truly hope that you never have an emergency and need to make use of this part. But if you do, it provides information that every person needs to know — for your own safety or to assist others.

Appendix A contains information about first aid techniques that can save lives until help arrives. Appendix B is a fast finder, listing first aid treatment to use, depending on the creature encountered.

Appendix A

First Aid Techniques

*I*n this appendix, we present three life-saving procedures:

- ✔ Mouth-to-mouth resuscitation and external heart massage.
- ✔ What to do for anaphylactic shock.
- ✔ How to apply a pressure immobilisation bandage.

Life-threatening situations must always be treated first, because without breathing or heartbeat, body tissues begin dying. After regular breathing and blood circulation are established, you can treat the injury or illness.

Mouth-to-Mouth Resuscitation and External Heart Massage

When a victim collapses or appears unconscious, follow the **DRABC** (Danger, Response, Airway, Breathing, Circulation) action plan:

1. **Danger — check for hazards or risks to yourself or others before assisting the victim.**

2. **Response — look, feel and listen for signs of breathing and a heartbeat or pulse. Ask the victim to squeeze your hand or respond in some way.**

3. **Airway — check the victim's mouth and remove anything he or she could choke on — for example, vomit, broken teeth or water. See Figure A-1.**

   a. Empty the mouth by turning the victim on his or her side and gently turning the face downwards to allow drainage. It may be necessary to use your fingers to clear the mouth.

   b. Open the airway by placing the victim on his or her back and tilting the head back and the chin up.

4. **Breathing — look, listen and feel for breathing and begin mouth-to-mouth resuscitation if the victim isn't breathing.**

   Tilt the victim's head back and lift the chin to open the airway, then pinch the nose and cover your mouth over the victim's mouth and begin resuscitation by blowing two breaths, as shown in Figure A-1, one second apart.

   If the victim has suffered facial injuries, breaths can be given through the victim's nose while holding the mouth shut.

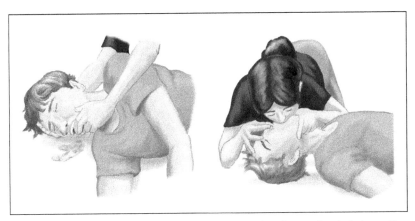

Figure A-1: How to perform mouth-to-mouth resuscitation. Ensure the victim's mouth is empty (left). Cover the victim's mouth with your mouth (right).

5. **Circulation — check for a pulse in the neck or for a heartbeat and, if absent, begin cardiac massage. See Figure A-2.**

   Kneel beside the patient and begin cardiac massage by depressing the lower half of the sternum (breast bone) one-third the depth of the chest for 30 depressions at a rate of 100 per minute (approximately two per second), followed by two breaths.

   (The Australian Resuscitation Council recommends this rate for all ages — infants, children and adults.)

6. **Call 000 for an ambulance.**

   The emergency services number throughout Australia is 000. If you dial 911, you're diverted to 000.

7. **Continue with mouth-to-mouth resuscitation and external heart massage until the victim recovers or medical assistance arrives.**

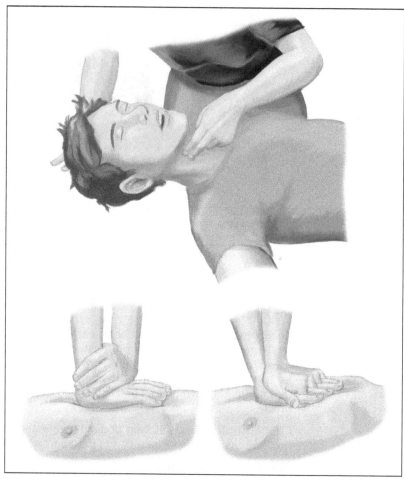

Figure A-2: How to give external heart massage. Checking for a pulse (top). Hand placement (bottom).

Anaphylaxis

Anaphylaxis (anaphylactic shock) is a sudden, severe allergic reaction to certain materials, substances or organisms.

Anaphylaxis involves various systems in the body (such as the skin, respiratory tract, digestive tract and cardiovascular system).

Severe symptoms or reactions to the allergen (food, bite or sting) that require immediate medical attention include: Difficulty breathing, swelling (particularly of the face, throat, lips and tongue, in cases of food allergies), rapid drop in blood pressure, dizziness, skin rash or hives, or unconsciousness.

Anaphylaxis can happen just seconds after being exposed to an allergen or can be delayed for up to two hours if the reaction is from a food. Fortunately, severe or life-threatening allergies occur in only a small group of people. Most anaphylactic reactions — up to 80 per cent — are caused by peanuts or tree nuts.

Here's what to do to treat anaphylactic shock:

1. **Put on disposable gloves (if available).**
2. **Call '000' for an ambulance.**
3. **Calm and reassure the victim.**
4. **Administer mouth-to-mouth resuscitation and external heart massage if the victim stops breathing and you can't find a pulse.**

   Refer to the steps outlined earlier in this chapter.

EpiPen administration

Adrenaline (epinephrine) is the only drug that acts fast enough to rescue someone from a life-threatening reaction. Victims who have a history of anaphylaxis often carry adrenaline in the form of an EpiPen for self-administration should the need arise.

In an emergency, you can assist the patient to self-administer EpiPen adrenaline by:

1. Removing the injector from the packaging.

2. Removing the safety cap.

3. Placing the EpiPen against the patient's mid thigh — no need to remove clothing.

4. Hold the EpiPen (firmly) with the tip at right angles against the thigh, but don't cover the end of the injector with your thumb!

5. Press hard. You should hear a click.

6. Apply moderate pressure and hold for 10 seconds.

7. Remove the EpiPen and rub the area for 10 seconds.

8. Remain vigilant for signs of a relapse — severe symptoms sometimes recur after apparent recovery.

9. Discard the EpiPen safely.

The Pressure Immobilisation Bandage

The pressure immobilisation bandage is the standard first aid method for:

- ✔ Venomous bites from snakes, funnel-web spiders and the blue-ringed octopus
- ✔ Cone shell injuries

The application of pressure to a venom-loaded bite, plus immobilisation of the limb, reduces the spread of venom and gives you more time to reach medical assistance before the victim suffers severe poisoning symptoms.

The pressure immobilisation bandage should *not* be used for:

- ✔ Ant, bee, Red-back Spider, White-tailed Spider, wasp and tick bites
- ✔ Jellyfish stings, including the Bluebottle
- ✔ Fish spikes or Platypus spur injuries

The pressure immobilisation technique

Here's how to apply a pressure immobilisation bandage and splint to the affected area:

1. **Apply a firm bandage.**

   Start by applying the bandage at the end of the limb, leaving tips exposed, then work upwards towards the knee or elbow. Bandage as much of the limb as possible. See Figure A-3.

   Crepe bandages or pantyhose work equally well. Note that you don't need to remove clothing, but thick seams should be cut away or pushed aside.

2. **Immobilise the limb with a splint.**

   Place a splint beside the injured limb and bind it to the limb with another bandage or string. See Figure A-3.

   Any straight rigid object can be used as a splint — wood, rolled-up newspaper or clothing.

3. **Keep the victim calm and still.**

4. **Transport the victim to a hospital.**

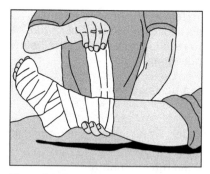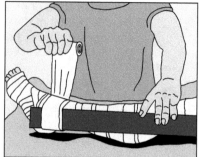

Figure A-3: Applying a pressure immobilisation bandage. Bandaging (left). Applying a splint (right).

Exceptions

Fortunately bites to the face and body (aside from the limbs) are uncommon. But, should this be the case, place a wad of cloth firmly over the bite and seek medical attention.

Infection . . . days later

An injury can easily become infected if it's not thoroughly clean, and a few days later it can become serious or even life-threatening. Signs of infection are redness, swelling, weeping from the wound, pus, increased pain, or a redness spreading from the wound area to other parts of the limb or body. Seek medical treatment to combat the infection.

Appendix B

First Aid Fast Finder

*I*n this appendix we provide first aid information on the animals you're most likely to encounter at home, at the beach or in country Australia. Cross references to chapters are provided throughout this appendix, so that you can find out more about the animal that has caused the injury.

First aid is immediate emergency care or assistance. Carry out the appropriate treatment and then get medical help from qualified paramedics or medical staff if necessary.

Bats

To treat a bite or a scratch, wash the wound thoroughly with soap and warm water for five minutes, then contact a doctor immediately (for a rabies vaccination). To remove bat saliva from a victim's eyes, nose or mouth, flush thoroughly with water, then seek medical advice. Refer to Chapter 7.

Birds

To treat minor beak wounds, clean the affected area thoroughly with soap or an antibacterial solution and warm water, then bandage. To treat talon wounds, clean with antiseptic solution or soap and apply antibiotic ointment before bandaging. Monitor wounds and consult a doctor if signs of infection occur. For more information on bird attacks, refer to Chapters 8 and 10.

Seek urgent medical assistance in the event of eye injuries, a suspected bone fracture or serious lacerations. Apply pressure to serious lacerations to stem blood flow until help arrives.

Blue-ringed octopuses

Call an ambulance. Administer mouth-to-mouth resuscitation at the first sign of respiratory failure and continue until medical help arrives. Refer to Chapter 11 for more information.

Bristle worms

If the bite site or the victim's skin becomes inflamed or irritated, consult a doctor. Refer to Chapter 11.

Cane Toads

To remove toad venom from a victim's eyes, mouth, face or any other area on the body, douse with plenty of water. Seek medical attention if stinging persists. Refer to Chapter 15.

Caterpillars

Apply a cold pack to reduce local pain and swelling. If embedded hairs can't be washed or stripped off, or if irritation continues for a long period, seek medical assistance. Refer to Chapter 14 for more details.

Centipedes

Apply a cold pack or take a generic painkiller to reduce pain. If dizziness, headache, nausea and/or vomiting occur, seek urgent medical attention. Refer to Chapter 6.

Cone shells

Seek urgent medical attention. Apply a pressure immobilisation bandage. If the victim becomes unconscious, administer mouth-to-mouth resuscitation and external heart massage (refer to Appendix A) until help arrives. For more information on cone shells, refer to Chapter 11.

Coral

For a minor cut or graze, wash with an antiseptic solution and cover with a bandage. For severe wounds or cuts, rinse with vinegar to help dissolve and remove coral dust, then rinse a second time with clean water, then apply a bandage. Check the wound daily for signs of infection. Refer to Chapter 11 for more details.

Crown of Thorns Starfish

Pull out spines that can be easily removed. Seek medical assistance to remove deeply embedded spines. Use very warm water to dull the pain and disperse the venom, then rinse and clean area with sterile water. Apply antibiotic ointment (if available) and a clean bandage. Dress wound daily and check for signs of infection. Refer to Chapter 11 for more information about the Crown of Thorns Starfish.

If the victim displays an allergic reaction — rapid swelling and intense pain around the wound site — seek urgent medical assistance.

Crustaceans (crabs, lobsters, yabbies, prawns)

Bites usually cause bruising only, but if the skin is broken, bathe the wound with an antiseptic solution and apply a bandage. Check the wound daily for signs of infection. Refer to Chapter 11.

Crocodiles

Remove the victim from the water and apply pressure to the wound with a rolled-up towel (or whatever is at hand) to stem the bleeding. Raise an injured limb to reduce blood loss. Call for an ambulance immediately.

Avoid moving the patient while waiting for the ambulance — movement increases blood loss. Keep the victim as calm and warm as possible. For more information on crocodile attacks, refer to Chapter 4.

Dingoes

If the wound is serious, apply pressure to the area to stem blood flow and call an ambulance. To treat minor wounds, clean the affected area thoroughly with soap or an antibacterial solution and warm water to remove dirt, saliva and bacteria. Allow the water to run over the wound for several minutes to ensure it's clean, then dry and bandage. Consult a doctor about a tetanus shot or if signs of infection occur. For more information on Dingoes, refer to Chapter 7.

Echidnas

Thoroughly clean spike wounds with an antibacterial solution and very warm water. Dry and bandage the wound. Check daily for signs of infection. Refer to Chapter 9 for more information on echidnas.

Electric Rays

Remove the victim from the water. Keep the victim calm and warm, and seek medical attention. In severe cases, the victim may stop breathing. Call an ambulance, and administer mouth-to-mouth resuscitation and external heart massage until help arrives. For more information on the Electric Ray (also known as the Torpedo Ray), refer to Chapter 12.

Feral animals (cats, dogs, cattle, goats, donkeys, horses, pigs)

If the wound is serious, apply pressure to stem blood flow and seek urgent medical assistance. To treat minor wounds, clean the affected area thoroughly with soap or an antibacterial solution and warm water to remove dirt, saliva and bacteria. Allow the water to run over the wound for several minutes to make sure it's clean, then dry and bandage. Consult a doctor about a tetanus shot or if signs of infection occur. For more information on feral animals, refer to Chapter 7.

Food poisoning

Pufferfish consumption: Call an ambulance. If respiratory failure develops, administer mouth-to-mouth resuscitation and external heart massage until medical help arrives. Refer to Chapter 10 for more information about pufferfish poisoning.

Other fish or crustaceans (crabs, lobsters, yabbies and prawns): Seek urgent medical attention. If the victim is unconscious, monitor breathing and administer mouth-to-mouth resuscitation and external heart massage, if necessary, (refer to Appendix A) until help arrives.

Insect bites

Ants: Apply a cold pack to reduce local pain and swelling. If the victim displays signs of an allergic reaction — breathing difficulties, a rash, facial swelling, vomiting, dizziness or unconsciousness — seek urgent medical help. To treat a severe allergic reaction until help arrives, refer to Appendix A. For more information on ants, check out Chapter 14. For more on Fire Ants, refer to Chapter 15.

Assassin bugs: Apply a cold pack to reduce local pain and swelling. Refer to Chapter 14.

Bedbugs: Wash bites with antiseptic solution and monitor the victim for signs of allergic reaction (refer to Appendix A). Check wounds daily for signs of infection. Refer to Chapter 15 for more information on bedbugs.

Bees: Remove the bee sting immediately either by scraping it off or quickly pulling it out. Apply a cold pack to relieve pain. If the victim displays signs of an allergic reaction (swelling of the tongue or throat, breathing difficulties, stomach cramps, vomiting, diarrhoea or unconsciousness), call an ambulance. Refer to Appendix A to treat life-threatening symptoms until help arrives. For more information on bees, refer to Chapter 14.

Fleas: Apply a soothing antiseptic cream and avoid scratching the affected area. If irritation is severe or continues for more than two or three days, seek medical attention. For more information on fleas, refer to Chapter 15.

Flies: Apply a cold pack and antiseptic lotion to relieve symptoms. Avoid scratching the affected area. If the victim displays more severe reactions, seek medical advice. For more information on flies, refer to Chapter 10.

Lice: Visit a chemist or pharmacy for a shampoo to treat Head Lice or Pubic Lice. No prescription is necessary. See a doctor for a prescription to purchase a product to treat Human Body Lice. For more information, refer to Chapter 15.

Mosquitoes: Wash bites with an antiseptic solution and apply a cold pack to help relieve pain and swelling. Avoid scratching bites; use calamine lotion or an insect-bite cream or spray (available at pharmacies) to help ease itchiness. Call an ambulance if bites develop into large hives or red swellings, or if the victim has breathing difficulties or is experiencing severe pain. For more information on mosquitoes and the different diseases they can transmit, refer to Chapter 6.

Wasps: Apply a cold pack to reduce local pain and swelling. If the victim displays signs of an allergic reaction (swelling of the tongue or throat, breathing difficulties, stomach cramps, vomiting, diarrhoea or unconsciousness), call an ambulance. Refer to Appendix A to treat life-threatening symptoms until help arrives. For more information on wasps, turn to Chapters 14 and 15.

Jellyfish (including Bluebottles)

Australian Box Jellyfish: As soon as possible, pour vinegar over the tentacles to deactivate them, soaking the area for at least 30 seconds. Only then can you attempt to remove the tentacles, very carefully, to ensure that no more venom is released. Wipe off inactive tentacles with a clean cloth or towel — don't rub. For more information on the Australian Box Jellyfish, check out Chapter 10.

Don't use methylated spirits, ammonia, urine or bicarbonate soda in place of vinegar.

If the victim loses consciousness, call an ambulance and perform immediate mouth-to-mouth resuscitation and external cardiac massage (refer to Appendix A), and continue until help arrives.

To relieve pain and help ease swelling in mild cases, use cold packs, painkillers and antihistamines.

Bluebottles: Remove tentacles with the pads of your fingers as soon as possible to stop stinging cells from 'firing' and releasing more venom. Soak area in very warm water to reduce pain and swelling. Painkillers (paracetamol) can also help reduce pain. (Other methods of pain relief are under investigation.) Refer to Chapter 10.

Chiropsalmus and Jimble jellyfish: Use vinegar to deactivate tentacles before removing them from the skin. Refer to Chapter 10.

Irukandji: Wash the area with vinegar to reduce pain (alternatives to vinegar are currently under investigation). Monitor the victim and provide reassurance.

Call an ambulance if the victim develops more serious symptoms, such as lower back pain, severe aches and shooting pains across the back and chest, muscle cramps in arms and legs, or headache, sweating, nausea or anxiety. For more information on Irukandji jellyfish, refer to Chapter 10.

Lion's Mane and Mauve Stinger: Wash tentacles with sea water and gently detach from the skin. Soak area in very warm water to reduce pain. Painkillers (paracetamol) can also help relieve pain. Refer to Chapter 10.

Kangaroos

Seek urgent medical assistance in the event of injuries to the trunk, or to the head or face to check for concussion. Apply pressure to serious wounds to stem blood flow, and raise if wound is on a limb, until help arrives. To treat minor cuts and scratches, wash the area with soap and water, then bandage. Monitor for signs of infection. Refer to Chapter 9.

Koalas

Thoroughly clean wounds with an antibacterial solution and very warm water, then dry and bandage. Refer to Chapter 9 for more information. Check daily for signs of infection (refer to Appendix A). A large or deep wound may need stitching; consult a doctor.

Leeches

Don't attempt to pull the leech off. Remove the leech by sprinkling salt, sea water or vinegar on it, or by applying a recently extinguished match to its body. Wash the wound with water and apply pressure to stem blood flow. Apply a cold pack to reduce pain or swelling, and check for signs of infection over the following days. Note that delayed irritation and itching can also occur. Refer to Chapter 6.

Millipedes

Flush the affected area thoroughly with lots of water and apply a cold pack to relieve itchiness and swelling. Seek medical attention if swelling spreads or pain persists. If the victim's eyes are affected, flush well with water and consult a doctor immediately. Refer to Chapter 6.

Numbfish

A shock from an electric Numbfish usually produces a temporary tingling sensation, similar to receiving a shock from an electric fence. Normally, no action is required. However, if the victim has a pre-existing heart condition, seek medical advice. Refer to Chapter 10.

Platypuses

If spurred, don't apply a cold pack to the wound — it only intensifies the pain and discomfort. If bleeding is heavy, a firm bandage can be applied. Call an ambulance and keep the victim calm until help arrives. For more information on Platypus spurs, refer to Chapter 9.

Possums

Clean the wound thoroughly with soap or an antibacterial solution and very warm water to remove dirt, saliva and bacteria. Dry and bandage the injury and check daily for signs of infection. Refer to Chapter 9 for more information on possums.

Rats

Thoroughly cleanse the wound using an antiseptic solution and monitor for signs of infection. Consult a doctor about a tetanus injection. For more information on rats, refer to Chapter 15.

Scorpions

Apply a cold pack to the affected area to help relieve pain. Painkillers, such as paracetamol, may also be helpful. If the victim develops an allergic response, such as breathing difficulties or goes into shock (refer to Appendix A), call an ambulance immediately. For more information on scorpions, refer to Chapter 6.

Sea Lizards

Wash affected area with salt water and gently detach tentacles. Soak area in very warm water to reduce pain. Painkillers (for example, paracetamol) can also help relieve pain. Refer also to Chapter 10.

Sea urchins

Carefully remove any spines protruding from the skin and bathe the wound in very warm water. If spines are deeply embedded, seek medical assistance. Wash with an antiseptic solution, then apply a bandage. Check daily for signs of infection. Refer to Chapter 11.

Sea snakes

Don't cut or apply suction to the bite. Avoid movement and apply a pressure immobilisation bandage to the limb (refer to Appendix A). At the first sign of any symptoms, call an ambulance. Be prepared to administer mouth-to-mouth resuscitation and external heart massage if the victim develops breathing difficulties. Refer to Chapter 11 for more information.

Sharks

Remove the victim from the water and call for an ambulance immediately. Apply pressure to the wound, using a rolled-up towel (or whatever is at hand) to stem blood flow. Raise the limb to reduce further blood loss. If the victim is wearing a wetsuit, don't remove it. Avoid moving the patient until help arrives. Refer to Chapters 10 and 12 for more information on shark attacks.

Snakes

Don't wash the area around the bite — the venom left on the skin can help doctors identify the snake and select the correct antivenom. Apply a pressure immobilisation bandage immediately (refer to Appendix A) and seek urgent medical attention. Keep the victim calm until help arrives.

Don't cut the area around the bite to make it bleed. Don't attempt to suck out the venom. Refer to Chapter 5 for more information.

Spiders

Collect the spider for ID purposes, if possible. Refer to Chapter 13 for information on spider species and characteristics, and additional first aid advice.

Red-back Spider: Apply a cold pack to the site of the puncture wound to reduce pain and swelling. Do not apply a pressure immobilisation bandage. Call an ambulance and keep the victim calm until help arrives.

Funnel-web spider and Mouse Spider: If the bite is on a limb, apply a pressure immobilisation bandage (refer to Appendix A) and call an ambulance. Restrict the victim's movement to slow down the spread of venom. Don't wash or wipe off venom present on the skin (it can be used by doctors to identify the spider and administer the correct antivenom). Keep the victim calm until help arrives.

Other spiders: Apply a cold pack to the site of the puncture wound to relieve pain and swelling. No further first aid treatment is required unless illness develops.

Spiky fish

Remove any broken pieces of spine embedded in the skin. Wash the injury in very warm water to help neutralise the venom and relieve pain. Generic painkillers (for example, paracetamol) can also be used to help relieve pain. Clean and bandage the injury. Check the wound daily for signs of infection. If symptoms other than pain occur, seek urgent medical assistance. Refer to Chapters 10, 11 and 12 for more information on different species of spiky fish.

Stingrays

Immerse the wound in very warm water to relieve pain and help neutralise the venom. Try to remove any fragments of the barbed spine from the wound, but only if this doesn't cause further damage or bleeding. Don't remove fragments sustained to the chest or abdomen. Do not use a pressure immobilisation bandage or try to close the wound, but apply pressure to bleeding wounds in the case of arterial damage. Seek medical treatment as soon as possible. Refer to Chapter 12 for more details.

Tasmanian Devils

Clean the wound thoroughly with an antibacterial solution and warm water to remove dirt, saliva and bacteria. Repeat twice to ensure all foreign matter is removed. Dry and bandage. If the wound is serious, seek urgent medical assistance. Consult a doctor about a tetanus shot or if signs of infection occur. For more information on Tasmanian Devils, refer to Chapter 7.

Ticks

Use tweezers to remove a tick as soon as it's detected. Refer to Chapter 6 for instructions. Keep the tick in a glass jar for 4 days (for ID purposes) and monitor the patient during this time for signs of an allergic reaction. To deal with an infestation of larval ticks, dissolve 1 cup (250 grams) of bicarbonate of soda in a warm bath and soak in it for 30 minutes.

If the victim faints, turns pale and develops breathing difficulties, call an ambulance immediately and administer mouth-to-mouth resuscitation and external heart massage until help arrives. Refer to Appendix A for more information on anaphylactic shock.

Water buffaloes

Stem bleeding by applying pressure with a towel (or whatever is at hand). If the injury is on an arm or a leg, raise the limb to slow blood loss. Call an ambulance. Refer to Chapter 7.

Index

. . . .

Notes

Notes

Australia's Dangerous Creatures For Dummies®

BESTSELLING
BOOK SERIES

Where the Deadliest Creatures Live

- **Eastern (Common) Brown Snake:** Eastern and central Australia, especially during the day in summer months. Beware of long grass.

- **Eastern (Mainland) Tiger Snake:** Victoria, eastern New South Wales and a tiny area of southeast Queensland, near rivers, creeks, lakes and dams. Often visits farms and outer suburban homes. Active during the day and warm, humid nights.

- **Coastal Taipan:** On and near the eastern coast from northern New South Wales and across the north coast of Australia, as far west as the top end of Western Australia. Also present on Fraser Island, Queensland. Often visits farm buildings, rubbish tips and sugarcane fields.

- **Death adder (four species):** All mainland states and territories, except Victoria. Usually concealed in fallen leaves, sand or gravel, grass and other ground cover. More active at night.

- **Estuarine (Saltwater) Crocodile:** North of the Tropic of Capricorn, in estuaries, along beaches and in the open sea. Also in freshwater billabongs, creeks and swamps up to 200 kilometres (124 miles) inland.

- **Dingo:** Farming and outback areas across most of mainland Australia, except for agricultural areas in Victoria and New South Wales. Also on Fraser Island, near Brisbane, Queensland.

- **Great White Shark:** In temperate waters along the coasts of New South Wales, Victoria, Tasmania, South Australia and the southern part of Western Australia.

- **Tiger Shark:** In shallow tropical and subtropical waters along the north, west and east coasts.

- **Bull Shark:** In tropical and subtropical waters as far south as Perth on the west coast, and Sydney on the east coast of Australia. Also found in rivers and estuaries — even in freshwater.

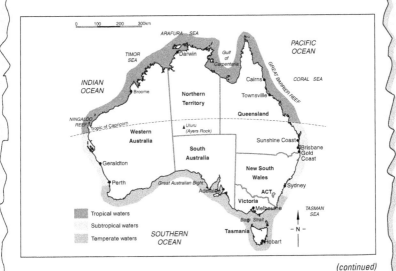

(continued)

For Dummies®: Bestselling Book Series for Beginners

Australia's Dangerous Creatures For Dummies®

Cheat Sheet

(continued)

- **Reef Stonefish:** Tropical coastal waters north of the Queensland Gold Coast and Geraldton in Western Australia. Lives near the bottom of coral and rocky reefs, often hiding under rocks, ledges, sand or mud.
- **Cone shell:** Along the northern, eastern and western coasts — but no further south than Sydney and Perth. Lives in mud and sand flats, rock pools and shallow reef waters, particularly at low tide.
- **Australian Box Jellyfish:** In shallow waters, estuaries and creeks along the northern half of the Australian coast, especially after rain.
- **Irukandji jellyfish:** In tropical coastal waters around the northern half of Australia. Small enough to get through holes in 'stinger' nets.
- **Bluebottle:** Carried by ocean currents and wind onto beaches, along the entire Australian coastline, including Tasmania.
- **Paper wasp:** Throughout Australia. Nests in the foliage of trees and under the eaves and verandas of rural and suburban homes.
- **European Wasp:** The cooler, damper parts of southern and eastern Australia, including urban areas. Nests underground or in cavities in walls or in trees.
- **Honey Bee:** Throughout Australia, including suburban backyards, parks and gardens.
- **Bull ant:** Throughout Australia in large nests several metres below ground. Nests have small entrances at surface level.
- **Jumping Jack ant:** South Australia, Victoria, New South Wales, Tasmania and the southern parts of Queensland and Western Australia.
- **Paralysis Tick:** Inhabits the 20–30-kilometre (12–19-mile) stretch of land along the eastern coastline of Australia. Prefers humid, moist and well-vegetated areas. Carried from place to place on the bodies of warm-blooded animals.
- **Red-back Spider:** Urban environments throughout Australia, but rarely in the bush. Found in dry and sheltered areas — wood-piles, garden sheds and the undersides of tables and chairs.
- **Sydney Funnel-web Spider:** Within a 120-kilometre radius of Sydney, usually in suburban gardens. Nests in rock crevices, stone walls, logs and in the ground.

Emergency Contacts

Ambulance, police, fire authorities: 000

Poisons Information Centres: 13 11 26

Roadside assistance (accidents, breakdowns and injured wildlife): 13 11 11

Travel Checklist

- Always carry a first aid kit appropriate to your needs
- Ensure you have good maps of the area
- Carry plenty of water and sun protection
- Have a means of communication with you (mobile phone, flares, mirror)

For Dummies®: Bestselling Book Series for Beginners

CPSIA information can be obtained
at www.ICGtesting.com
Printed in the USA
FSHW020845290519
58517FS